Data-Driven Methods for Dynamic Systems

Data-Driven Methods for Dynamic Systems

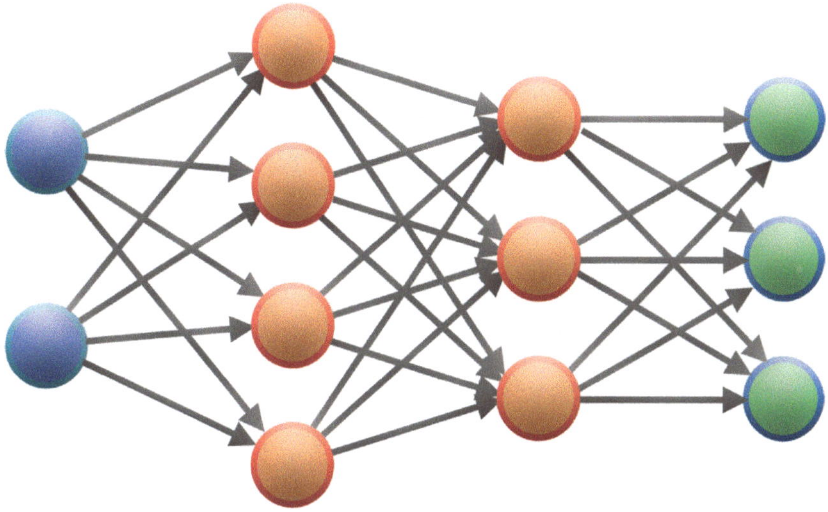

Jason J. Bramburger
Concordia University
Montréal, Québec
Canada

siam.
Society for Industrial and Applied Mathematics
Philadelphia

10 9 8 7 6 5 4 3 2 1

Publications Director	Kivmars H. Bowling
Executive Editor	Elizabeth Greenspan
Acquisitions Editor	Paula Callaghan
Developmental Editor	Rose Kolassiba
Managing Editor	Kelly Thomas
Production Editor	Ann Manning Allen
Copy Editor	Bruce Owens
Production Manager	Rachel Ginder
Production Coordinator	Cally A. Shrader
Compositor	Cheryl Hufnagle
Graphic Designer	Doug Smock

Library of Congress Cataloging-in-Publication Data

Names: Bramburger, Jason J., author.
Title: Data-driven methods for dynamic systems / Jason J. Bramburger.
Description: Philadelphia : Society for Industrial and Applied Mathematics, [2024] | Series: Other titles in applied mathematics ; 201 | Includes bibliographical references and index. | Summary: "Brings together modern computational tools to provide an accurate understanding of dynamic data"-- Provided by publisher.
Identifiers: LCCN 2024027823 (print) | LCCN 2024027824 (ebook) | ISBN 9781611978155 (paperback) | ISBN 9781611978162 (ebook)
Subjects: LCSH: System analysis. | Differential equations. | AMS: Ordinary differential equations -- Differential equations in the complex domain -- None of the above, but in this section.
Classification: LCC QA402 .B637 2024 (print) | LCC QA402 (ebook) | DDC 003/.85--dc23/eng20241001
LC record available at *https://lccn.loc.gov/2024027823*
LC ebook record available at *https://lccn.loc.gov/2024027824*

Contents

List of Algorithms

Preface

This book grew out of multiple stimulating conversations with Nathan Kutz while I was a postdoc at the University of Washington. It began with joking about taking our favorite dynamical systems textbook, *Nonlinear Oscillations, Dynamical Systems, and Bifurcations of Vector Fields* by J. Guckenheimer and P. Holmes, and rewriting it chapter by chapter with modern data-driven techniques. Our idea was to highlight how so much of the dynamical systems theory we were taught can now be explicitly implemented using the widely available computational techniques that typically fall under the umbrella of data analysis. Ideas as simple as optimally fitting data to models could lead the reader right back to Guckenheimer and Holmes's textbook since one is now in a position to apply all the pencil-and-paper techniques that have been developed over more than a century of dynamical systems theory. Similarly, finding changes of variable that recast complex dynamics in the form of the phenomenological models that have been examined in detail by not only Guckenheimer and Holmes but also nearly anyone who has taught or published in dynamical systems is now accessible using neural networks. The goal has always been to showcase a suite of computational methods that can be combined with analysis to better understand the increasingly complicated models and datasets that describe our complex world.

I began taking the idea of writing a textbook on data-driven methods for dynamical systems seriously early in the spring of 2022. Like most of the world, I found myself inside due to a global pandemic with an abundance of time to read new papers and write my own. This textbook started with me compiling notes on interesting things I was reading. I further began documenting both the successes and the failures of these data-driven methods to better understand the systems I was interested in. The greatest leap forward in transforming my notes into this textbook came from weekly meetings between myself and the mathematical data science group at the University of Surrey: Stefan Klus, David Lloyd, and Naratip Santitissadeekorn. At the time I was a fellow of the Institute of Advanced Studies in Guildford, UK, and collaborating on projects in data science and pattern formation with Stefan, David, and Naratip. These conversations introduced me to new methods and techniques that convinced me that they should be documented in one place for the ease of reference. I am eternally grateful for the mentorship and guidance offered to me by these three excellent mathematicians. It was particularly Stefan Klus's work with kernel methods that moved me beyond my (at that time) limited purview of data-driven methods for dynamical systems and convinced me that the field has been developed enough to necessitate a textbook.

An early version of this textbook was used as the basis for the notes in my *Data-Driven Methods for Dynamic Systems* course held in the winter semester of 2023 at Concordia University. The goal of this course was threefold: to share my passion for dynamical systems with the students, to provide students the opportunity to simultaneously learn advanced dynamical systems theory while implementing the theory computationally, and for me to finalize and fine-tune the organization and presentation of the text. The accompanying scripts and notebooks for this textbook were created initially as demonstrations for the students to provide a foundation for their own work in the course assignments. These computational implementations are freely available for anyone to access (see **https://bookstore.siam.org/ot201/bonus**) and should be viewed as the supplementary material accompanying this book. These computational implementations can be used to reproduce the results presented in the examples and to form a basis for the application of the methods to other systems or data of interest. Thus, as was my intention for my students, these scripts and notebooks guarantee that the reader does not need to start from scratch should they wish to implement the methods outlined in what follows.

In writing this book I have carefully curated methods that I believed to be at the forefront of computational mathematics and that can precisely aid or complement traditional dynamical systems analysis. This means that I have neglected certain methods, such as diffusion maps, since from my perspective it has fallen out of favor by the community and has been replaced with more robust machine learning approaches that are covered in the book. Similarly, the emerging methods surrounding neural differential equations have been omitted since they are primarily for forecasting dynamical systems. While interesting in their own right, forecasting dynamics is slightly away from the focus of this book as it does not rely on exploiting any dynamical features of the system, making it more closely related to numerical analysis than dynamical systems theory. That is, this book is mostly concerned with interpreting data from a dynamical systems perspective, meaning classifying features such as fixed points, periodic orbits, chaotic attractors, invariant measures, and so on.

One should note that this is not a textbook in the traditional undergraduate text definition. From my perspective, this book should function more as a gentle introduction to the field of data-driven methods for graduate students and even established researchers. That is, it should be seen as a starting point for familiarizing oneself with these methods, while references are made throughout to the source material where greater detail can be found. This therefore provides the reader the opportunity to dig into the details should they require it while also keeping the presentation here relatively light and exemplary. One should be warned that the presentation assumes a basic familiarity with dynamical systems theory, with a required foundation of at least an undergraduate course in the subject. Furthermore, beyond the examples presented in each section, one will not find any exercises herein. The reason for this is simple: They are hard to come up with. The work covered in this book is quite new, and so there is not a wealth of "simple" problems that students can work through in a timely manner. While teaching the aforementioned course based on this material, I provided long, open-ended problems for students to apply these methods. The timeline for each project was one full month at a time with a full page of explanation for what they could try and should look for. Therefore, this book should act as a first point of contact for students and faculty in applied and computational mathematics, as well as engineers and computer scientists, who wish to be informed on these methods without having to dig through mountains of papers or conference proceedings themselves.

Finally, beyond those mentioned above, a serious debt of gratitude is owed to Steven Brunton. He was my co-mentor with Nathan Kutz at the University of Washington, and it was his ambition and curiosity that really stimulated my curiosity in many of these methods beyond my previously limited interests. Steve's ability to communicate complex ideas in a simple way demonstrated a mastery of the subject that I have always been envious of. Much of this book attempts to channel this ability by providing illustrative examples that communicate the major concepts that demonstrate the potential of the methods while pointing the reader to the source material should one wish for a more technical presentation.

Chapter 1

Dynamical Systems: Old and New

With the growing size of experimental data and the increase of computational power, there have been revolutionary new tools developed to model these datasets. Broadly falling under data science and machine learning, these techniques have the power to not only model new systems but also fundamentally alter how we analyze the systems we already have. In this book we will bring together computational tools such as neural networks, sparse regression, dynamic mode decomposition, and semidefinite programming to provide an accurate understanding of dynamic data. These techniques build upon the extensive work from pencil-and-paper dynamical systems techniques that go back decades and sometimes even centuries. The result will be an introduction to methods that complement, not replace, the analysis of time-dependent systems.

1.1 ▪ What Is Dynamical Systems Analysis?

Throughout this book we will consider *dynamic* data and take a dynamical systems approach to understanding it. Generally, this means that we are interested in systems that change in time. As mathematical models, dynamic systems can arise in the form of ordinary and partial differential equations, Markov chains and stochastic differential equations, difference equations, or even cellular automata. The change in time might be discrete, representing a hopping from one instance to the next, or continuous, flowing through time moment by moment. So long as there is a component of temporal change, one can reasonably refer to a model as a **dynamical system**.

So, it is reasonable to ask what exactly dynamical systems analysis is and what differentiates it from other forms of mathematical investigation. That is, does the presence of a time derivative or an iteration scheme in the model you've chosen to investigate necessarily imply that you're doing dynamical system analysis? As is natural with breaking up any subject into distinct branches, the dividing lines are ill-defined and the subject of open debate. Nonetheless, what typically differentiates dynamical systems as a discipline from just analyzing time-dependent systems is the questions asked about the system and the methods used to arrive at the answers. Precisely, dynamical systems analysis emphasizes a geometric interpretation of the system. Classical analysis of differential equations at least starts with existence and uniqueness theorems, while the geometric intuition of dynamical systems might emphasize the existence of invariant manifolds in phase space; see Figure 1.1 for a comparison.

For demonstration, let's consider E. N. Lorenz's famous ordinary differential equation (ODE) [178]

$$\begin{aligned} \dot{x}_1 &= -\sigma x_1 + \sigma x_2, \\ \dot{x}_2 &= \rho x_1 - x_2 - x_1 x_3, \\ \dot{x}_3 &= -\beta x_3 + x_1 x_2, \end{aligned} \tag{1.1}$$

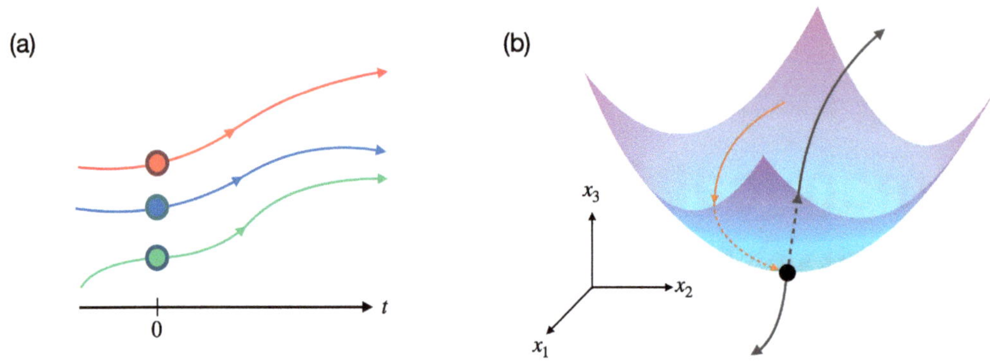

Figure 1.1. *Dynamical systems analysis takes a geometric approach to understanding the evolution of time-dependent trajectories evolving through space.* (a) *Standard existence and uniqueness results are a critical first step that tell us that solutions exist locally in time and that solutions that start close remain close over finite timescales.* (b) *The stable manifold theorem gives the existence of flow-invariant stable (colored two-dimensional manifold) and unstable (gray one-dimensional curve) manifolds containing initial conditions whose trajectories converge to a steady state (black dot) as $t \to \infty$ or $t \to -\infty$, respectively. Arrows indicate the direction of movement as t increases.*

where σ, ρ, and β are positive real-valued system parameters and the dots are Newton notation for the derivative with respect to time, t. A standard question that arises in both the study of ODEs and partial differential equations (PDEs) is to determine whether solutions exist and are unique for classes of initial conditions. Existence and uniqueness of solutions lies at the heart of the analysis of differential equations, while such critical results for many PDE models remain outstanding and are an area of ongoing research.[1] Returning to the Lorenz system (1.1), the polynomial nonlinearity of the right-hand side means that standard results guarantee that for any initial condition, (x_1^0, x_2^0, x_3^0), there is *exactly one* smooth trajectory $(x_1(t), x_2(t), x_3(t))$ satisfying the differential equation for all t in some interval about $t = 0$ with

$$(x_1(0), x_2(0), x_3(0)) = (x_1^0, x_2^0, x_3^0). \tag{1.2}$$

This can be extended to demonstrate continuous dependence of trajectories over finite time intervals with respect to initial conditions and system parameters. Further, one can prove that all trajectories, no matter where they start, will remain bounded for all $t \geq 0$, giving existence for all forward time. While this last point is unique to the Lorenz system, existence, uniqueness, and continuous dependence on parameters are now standard results in the theory of ODEs [282].

Existence and uniqueness of solutions provides the critical first step for understanding a dynamical system. For example, it tells an analyst that trajectories of (1.1) do not intersect in the three-dimensional phase space $\mathbb{R}^3$, i.e., the space of all possible initial conditions. However, the existence of a solution provides almost no information about how the solutions evolve. Many questions in dynamical systems revolve around how solutions evolve in phase space, particularly where they are going and how they get there. This is where the geometric perspective comes in. One is interested in manifolds embedded in phase space that are invariant with respect to the flow of a system and what potential limit sets look like.

Solutions that are of particular interest in dynamical systems are **steady states**, or **equilibrium solutions**. They are characterized by not changing in time and are identified by setting all time derivatives in the system to zero. Returning to the Lorenz system (1.1) for exposition, the origin $(x_1, x_2, x_3) = (0, 0, 0)$ is a steady-state solution. Thus, the initial condition

[1] At the time of writing this, existence and uniqueness of solutions to the three-dimensional Navier–Stokes equations is so important that there is a US $1 million bounty on a proof or counterexample [85].

$(x_1^0, x_2^0, x_3^0) = (0,0,0)$ guarantees that $x_1(t) = x_2(t) = x_3(t) = 0$ for all $t \in \mathbb{R}$. Although trivial, such a point represents a zero-dimensional invariant manifold containing solutions to the Lorenz system. Moreover, we may compute the Jacobian matrix of the right-hand side of the Lorenz system about the origin to find

$$\begin{bmatrix} -\sigma & \sigma & 0 \\ \rho & -1 & 0 \\ 0 & 0 & \beta \end{bmatrix}, \tag{1.3}$$

which has eigenvalues

$$-\beta, \quad -\frac{1}{2}(\sigma+1) \pm \frac{1}{2}\sqrt{1 - 2\sigma + \sigma^2 + 4\rho\sigma}. \tag{1.4}$$

For every $\sigma, \beta > 0$ and $\rho \in (0,1)$ these eigenvalues are negative, in which case the origin is said to be **stable**. The stable manifold theorem guarantees that when the origin is stable, all initial conditions that are chosen sufficiently close to the origin will result in trajectories satisfying $(x_1(t), x_2(t), x_3(t)) \to (0,0,0)$ as $t \to \infty$ [109]. Thus, we now have a portrait of the state-space geometry induced by the dynamics of (1.1) in a neighborhood of the origin.

For $\rho > 1$ we find that one of the eigenvalues (1.4) becomes positive, while the other two remain negative. The stable manifold theorem can again be applied to guarantee the existence of two smooth invariant manifolds living inside the phase space $\mathbb{R}^3$: the stable and the unstable manifolds. In this case the stable manifold is two-dimensional, coming from the two negative eigenvalues, and is tangent at the equilibrium to the span of the associated eigenvectors. All initial conditions chosen on the stable manifold will converge to the origin as $t \to \infty$. Similarly, the unstable manifold is one-dimensional, coming from the single positive eigenvalue, and this manifold is tangent at the origin to the eigenspace spanned by the associated eigenvalue. All initial conditions chosen on the unstable manifold will converge to the origin in backward time, i.e., as $t \to -\infty$. The result is a local splitting of the phase space $\mathbb{R}^3$ near the origin into two embedded manifolds for which the flow on them is entirely determined in either forward or backward time.

There are many well-developed techniques that provide methods to understand local and even global temporal evolution of a system. ODEs and finite-dimensional difference equations are accompanied by some of the most well-developed theory, as is demonstrated in the numerous textbooks written about dynamical systems analysis (see, for example, [109, 264, 162]). Extending many of these ideas to more complex systems such as PDEs and stochastic models is an ongoing area of research, with the goal remaining the same: to better understand the state-space geometry induced by the flow of a time-dependent system. Beyond the local stability results outlined here, this further includes identifying transformations to canonical forms, finding invariant measures that describe the probability of finding trajectories in certain regions of space, and classifying the types of invariant manifolds that may be present in a system. From just this list alone we find that dynamical systems incorporates ideas from not only geometry but also algebra (canonical forms), probability and ergodic theory (invariant measures), and topology (invariant manifolds), to only name a few. Therefore, dynamical systems analysis presents itself an application of what are sometimes viewed as the most abstract areas of mathematics.

1.2 ▪ The Influence of Computers

It is hard to overstate the influence that computers have had on dynamic systems. Early applications of modern computers were to solving differential equations related to ballistic trajectories. Their ability to perform complex calculations much faster than humans meant that systems could be more accurate through the incorporation of physical effects that would previously have made

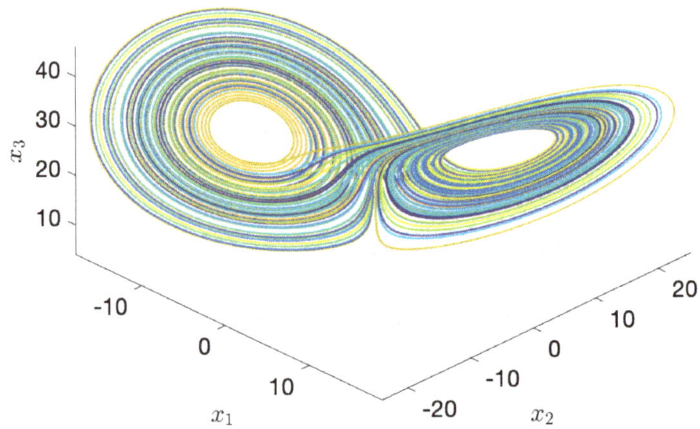

Figure 1.2. *The butterfly attractor of the Lorenz system* (1.1) *at the standard parameter values* $(\sigma, \beta, \rho) = (10, 8/3, 28)$. *Plotted is a single trajectory as it moves through* (x_1, x_2, x_3)*-space in time. Colors correspond to the value of* t*, with purple representing* $t = 0$ *and yellow representing* $t = 100$.

the equations too difficult to work with. Through their ability to handle arithmetic operations quickly, computers allowed for more intricate numerical methods that could reduce the error in approximating trajectories. This helped to overcome the fast error buildup leading to inaccurate predictions from the previously dominant Euler method while also lifting the tedious calculations from one timestep to the next out of the hands of the user.

Computers have made the theoretical achievable, at least approximately. With computers one is now able to approximate the long-time averages that form the basis of ergodic theory and its most important theorems. They allow one to generate (pseudo) random numbers to simulate stochastic processes. The simulation of PDEs has led to more aerodynamic motorized vehicles, thus resulting in greater fuel efficiency. At the base of this is the simple fact that computers have allowed us to visualize the complex motion of solutions to models in phase space. We are no longer forced to fixate on local regions of space, but computers enabled one to see the macroscopic, long-time behavior of a system. Such computation further drives forward the theory as well, as one can better understand what could be found in specific models.

The ability to "see" the trajectories of differential equations in phase space is no better typified than in the Lorenz system (1.1). With the help of E. Fetter and M. Hamilton, Lorenz simulated trajectories of his system to visualize their movement through the three-dimensional phase space. The results were staggering, as you can see for yourself in Figure 1.2. Lorenz and his team found that solutions evolved in a very complex manner, tracing out what is now known as the **butterfly attractor**. For the first time ever we were able to generate, view, inspect, and probe a chaotic system in the laboratory setting! Although it would take a decade or two to even formulate what it means to be chaotic, it was immediately noticed through just one figure, such as Figure 1.2, that computers could be used to go far beyond the theory that had been developed for centuries prior.

Seeing the butterfly attractor today may not seem special, as almost anyone reading this book will have grown up with computers. Indeed, complex graphics that are generated by simple rules, such as the butterfly attractor, are used to produce pieces of art [139] and make realistic graphics for video games [71]. Both of these advances are due to the computer revolution in dynamical systems, which goes at least back to the work of Lorenz.

Of course, computers offer us so much more than just visualization. They currently find themselves intertwined with the study of dynamical systems. Computations drive the theory forward while also validating the existing theory. Moreover, the computation itself is becoming an aspect of the theory. For example, in 1999, W. Tucker used a *computer-assisted proof* to provide the first demonstration that the Lorenz system (1.1) exhibits a robust (persists under perturbations) strange attractor [275]. Thus, computers not only offer a method of producing data but also offer a way of producing innovative methods that seek to address significant gaps between theory and the understanding of specific models through novel interrogations that draw inspiration from dynamical systems analysis.

1.3 ▪ Proper Orthogonal Decomposition

Topics from linear algebra, such as matrix decompositions, are some of the earliest applications of computational methods to be applied to better understand dynamic datasets. A perfect example of this is the **Proper Orthogonal Decomposition** (POD) [25], which represents a powerful melding of approaching old ideas in dynamical systems from a modern computational perspective. In this section we will review POD and demonstrate how it provides a way of decomposing space-time data via an optimal basis expansion that is obtained using only the singular value decomposition (see Appendix B). In this way, POD represents an early unification of simple computational methods with important analytical concepts from dynamical systems analysis. Moreover, POD is the same as principal component analysis in the field of statistics [133] and empirical orthogonal function analysis in signal processing [190], while performing the basis expansion is sometimes referred to as the Hotelling transform. The point is, POD is a widely used technique that has been discovered and rediscovered in a variety of scientific fields, all to do the same thing: better understand a dataset using easily implemented numerical methods.

We begin with the theory. In the analysis of space-time systems one often wishes to expand functions as the sum of some countable set of basis functions, $\{\phi_k(x)\}_{k=1}^\infty$, with time-dependent coefficients. The result is an expansion of the form

$$f(x,t) = \sum_{k=1}^\infty a_k(t)\phi_k(x), \tag{1.5}$$

where we notice that through this summation we have disentangled the space, x, and time, t, dependence in the system. Some examples that find application throughout mathematics are the Fourier expansion

$$f(x,t) = \frac{a_0(t)}{2} + \sum_{k=1}^\infty [a_k(t)\cos(kx) + b_k(t)\sin(kx)], \tag{1.6}$$

the Taylor expansion

$$f(x,t) = \sum_{k=0}^\infty a_k(t)x^k, \tag{1.7}$$

or wavelet series. Many of these expansions lie at the heart of the method of separation of variables in the analysis of PDEs [261] and thus have a history stretching back hundreds of years in mathematical analysis. In particular, with second-order linear PDEs in one spatial variable, the basis functions $\phi_k(x)$ can be obtained by solving an associated Sturm–Liouville problem, which are orthogonal with respect to an associated inner product. The utility of these basis expansions cannot be understated, with applications of them abounding throughout the analysis of differential equations, numerical analysis, image processing, and data science.

In practice, we often approximate a function by truncating the expansion (1.5) to a finite number of terms in the series. The result is a finite summation over a subset of the basis functions,

$$f(x,t) \approx \sum_{k=1}^{K} a_k(t)\phi_k(x), \tag{1.8}$$

with $K \geq 1$ chosen large enough to guarantee that the approximation is sufficiently accurate. This comes up with Taylor expansions when attempting to understand the flow of a dynamical system restricted to a small region of phase space, when using spectral methods to numerically approximate the solution to a PDE, and in taking the discrete Fourier transform of a signal to move to the frequency domain. There are many applications of series expansion in the form (1.5) that tell one exactly which basis functions to use to quantify the error in moving to a truncated sum. These applications are based sometimes on the ease of working with the basis functions, such as polynomials, or an intuitive understanding of what the basis expansions mean physically, such as using Fourier basis representations to move to the frequency domain.

Aside from the ease of analysis or mathematical intuition, there are other criteria one could choose to identify basis functions. One that is of practical and computational importance is identifying the basis functions so that we can best approximate a function with the fewest number of terms in the truncation (1.8), that is, which $\phi_k(x)$ allow one to use the smallest value of K to optimally approximate the function. The optimal choice of basis functions to achieve this are called the **proper orthogonal modes**, while expanding a function in these modes constitutes the aforementioned proper orthogonal decomposition. Knowing exactly what the proper orthogonal modes are is problem dependent, and there is no prescription for identifying them in general; however, we can use data to help approximate them.

Suppose we have a function $f(x,t)$, taking values in $\mathbb{R}^d$, that we wish to perform POD on. We assume that we have information about this function at $M \geq 1$ points in space, $\{x_1,\dots,x_M\}$, and $N \geq 1$ points in time, $\{t_1,\dots,t_N\}$. This information can be arranged in the data matrix

$$\mathcal{F} := \begin{bmatrix} f(x_1,t_1) & f(x_1,t_2) & \cdots & f(x_1,t_N) \\ f(x_2,t_1) & f(x_2,t_2) & \cdots & f(x_2,t_N) \\ \vdots & \vdots & \ddots & \vdots \\ f(x_M,t_1) & f(x_M,t_2) & \cdots & f(x_M,t_N) \end{bmatrix} \in \mathbb{R}^{(d\cdot M)\times N}, \tag{1.9}$$

where rows correspond to a position in space and columns correspond to points in time, termed **snapshots**. Applying the singular value decomposition (SVD) to this matrix results in orthogonal matrices $U \in \mathbb{R}^{(d\cdot M)\times(d\cdot M)}$ and $V \in \mathbb{R}^{N\times N}$, along with the diagonal matrix $\Sigma \in \mathbb{R}^{(d\cdot M)\times N}$. The columns of U approximate the proper orthogonal modes at each spatial coordinate $x_1,\dots,x_M$, while the columns of V approximate the temporal evolution of each mode at the time points $t_1,\dots,t_N$. This can be observed by writing $\mathcal{F} = U\Sigma V^*$ as the summation

$$\mathcal{F} = \sum_{k=1}^{\min\{d\cdot M,N\}} \sigma_k U_k V_k^*, \tag{1.10}$$

where U_k, V_k denote the kth column of matrices U and V, respectively, and σ_k is the kth singular value in the matrix Σ.

The SVD allows one to produce optimal low-rank approximations of the data matrix $\mathcal{F}$, which can in turn be used to produce optimal POD expansions with the fewest number of modes. Supposing we wish to use only r terms in our basis expansion, the SVD provides the optimal rank-r approximation of the data through

$$\mathcal{F}_r := \sum_{k=1}^{r} \sigma_k U_k V_k^*. \tag{1.11}$$

Algorithm 1.1 Proper Orthogonal Decomposition

1. Arrange data into matrix $\mathcal{F}$ of size (# space points) $\times$ (# time points)

2. Compute SVD $\mathcal{F} = U\Sigma V^*$

3. Proper orthogonal modes are given by columns of U with time evolution given by corresponding column of V

4. Fix $r \geq 1$ and construct rank-r approximation $\mathcal{F}_r := \sum_{k=1}^{r} \Sigma_{k,k} U_k V_k^*$

Moreover, one may further use the SVD to measure the quality of the approximation through the formula

$$\frac{\|\mathcal{F}_r\|_F^2}{\|\mathcal{F}\|_F^2} = \frac{\sigma_1^2 + \sigma_2^2 + \cdots + \sigma_r^2}{\sigma_1^2 + \sigma_2^2 + \cdots + \sigma_{\mathrm{rank}(\mathcal{F})}^2}. \tag{1.12}$$

This formula is sometimes referred to as capturing the *energy* ratio of the approximation. This terminology is motivated in part by the applications of POD to problems in physics, such as fluid flow [160, 25, 60, 117, 249]. We have that as r approaches the rank of the matrix $\mathcal{F}$, the ratio (1.12) approaches 1, representing capturing 100% of the original information encoded in $\mathcal{F}$. In many applications one is able to use a relatively small value of r and still achieve a ratio near enough to 1 to faithfully reconstruct the data, thus providing an enormous reduction in storing information contained in $\mathcal{F}$. Indeed, for a chosen value of r, one constructs $\mathcal{F}_r$ using $(d \cdot M) \cdot r + N \cdot r + r$ values, representing the first r columns of U and V and the first r singular values of Σ, as opposed to the full matrix $\mathcal{F}$, which contains $(d \cdot M) \cdot N$ values to store. The general POD algorithm is given in Algorithm 1.1 and illustrated with the following example.

Example 1.1. Let us consider the two-component reaction-diffusion PDE

$$\begin{aligned}\frac{\partial u}{\partial t} &= \frac{1}{10}\left(\frac{\partial^2 u}{\partial x^2} + \frac{\partial^2 u}{\partial y^2}\right) + u(1 - u^2 - v^2) + v(u^2 + v^2), \\ \frac{\partial v}{\partial t} &= \frac{1}{10}\left(\frac{\partial^2 v}{\partial x^2} + \frac{\partial^2 v}{\partial y^2}\right) - u(u^2 + v^2) + v(1 - u^2 - v^2).\end{aligned} \tag{1.13}$$

This PDE is well known to support spiral wave solutions [61], characterized by holding an Archimedean spiral profile in the u and v components while simply rotating about a fixed point in space as time evolves. In this example we will simulate one of these spiral waves to provide seed data to apply POD. They can be generated through the initial conditions

$$\begin{aligned}u_0(x,y) &= \tanh(\sqrt{x^2+y^2})\cos(\arctan(y/x) - \sqrt{x^2+y^2}), \\ v_0(x,y) &= \tanh(\sqrt{x^2+y^2})\sin(\arctan(y/x) - \sqrt{x^2+y^2}),\end{aligned} \tag{1.14}$$

which converge quickly into the true spiral wave solutions. The spatial domain is fixed as $[-10, 10] \times [-10, 10]$ with periodic boundary conditions. The spatial domain is discretized into a 256×256 grid, using spectral methods to transform (1.13) to the frequency domain, where it is integrated forward in time. We integrate from $t = 0$ to $t = 40$, collecting data at intervals of $h = 0.05$. In total, we have $d = 2$ coming from the two components u and v, $M = 256^2 = 65536$ coming from the 256×256 spatial grid, and $N = 801$ time points. The result is the space-time snapshot matrix $\mathcal{F}$ of size 131072×801, which contains 104988672 data points.

Figure 1.3(a) provides a snapshot of the spiral wave in the u component at $t = 40$, with v looking nearly identical. In panels (b) and (c) of Figure 1.3 the rank-1 and rank-2, respectively,

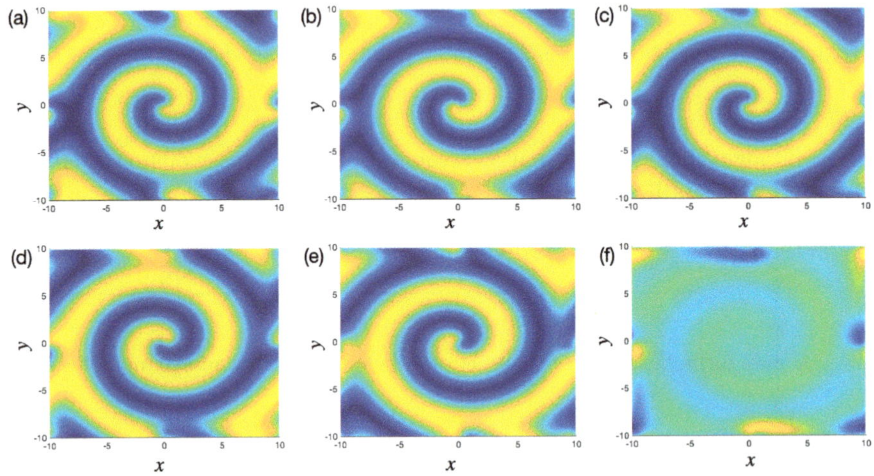

Figure 1.3. *A spiral wave and its POD reconstruction.* (a) *A snapshot of the* $u(x, y, t)$ *component of a spiral wave solution to* (1.13). (b) *A reconstruction of the spiral wave with only one POD mode and* (c) *two POD modes. The* (d) *first,* (e) *second, and* (f) *third POD modes.*

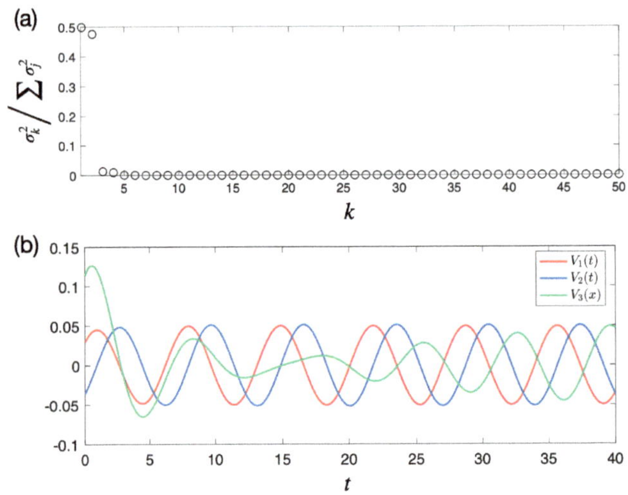

Figure 1.4. (a) *The amount of energy contained in each singular value of the spiral wave snapshots.* (b) *The first three columns of* V *in the SVD, giving the temporal evolution of the first three proper orthogonal modes.*

reconstructions of the u component of the spiral wave at $t = 40$ are provided as well. One is required to look quite closely to identify discrepancies between these low-rank reconstructions and the true snapshot in panel (a). We can see why this is the case from panels (d)–(f), which present the first three POD modes. The first two look almost identical to the original spiral wave, while the third appears to just fill in the details around the boundaries of the spatial domain.

We can further use the singular value spectrum and the formula (1.12) to quantify how faithfully the low-rank approximations reconstruct the data. Figure 1.4(a) provides the singular values plotted as the percentage of the total SVD energy. The first two singular values occupy more than 95% of the energy. This implies that one could faithfully reconstruct the space-time dynamics of the spiral wave snapshots in the data matrix $\mathcal{F}$ using only the first two columns of U and V and the first two singular values. This means storing only $2 \times 256^2 + 2 \times 801 + 2 = 263748$ data points, as opposed to the 104988672 values stored in $\mathcal{F}$. This represents a reduction by more than 99% of the stored data. Furthermore, the time periodicity of the dynamics coming from the rigid rotation of the spiral about a fixed point is entirely captured in the first two columns of V, as plotted in Figure 1.4. Notice that the third column of V is slightly more irregular, corresponding to the POD mode in panel (f) of Figure 1.3. This POD mode only makes up about 1.3% of the total energy and is therefore hardly noticeable if left out of a low-rank reconstruction. The third mode is likely an artifact of the convergence from the initial condition (1.14) into the true spiral wave solution and/or irregularities brought on by the boundary conditions.

POD is a perfect encapsulation of the topics to be covered throughout this book. It is well founded in the mathematical theory of basis expansions and functional analysis, with a history that stretches back centuries to some of the earliest pioneers in PDEs. This includes frequently encountered names in the theory of differential equations, such as Fourier, Lagrange, Bessel, Legendre, and the many others who now have basis functions named after them. The computational aspect of POD offers a new way of approaching an old topic. It uses matrix decompositions (another old idea from linear algebra) to provide a data-driven approach to identifying optimal basis expansions. The result is greater information about the coherent modes in the data and a way of storing and reconstructing it. It is one of the earliest data-driven methods that seeks to not only extract information from data but influence future analytical approaches as well.

1.4 ▪ What to Expect from This Book

The goal of this textbook is to provide an example-driven understanding of how modern computational tools can be used to interpret dynamic data. The methods will draw inspiration from problems and techniques in the theory of dynamical systems. Thus, one can expect sections to be organized in much the same way as the previous one on POD. We will start with a problem, theoretical technique, or concept from dynamical systems and then showcase how one may develop and implement computational methods based on it. Chapters will be organized around a central theme, while sections will seek to demonstrate a particular method and the theory it is evoking.

The techniques in this book build upon extensive work using mathematical analysis to interpret dynamic models, resulting in methods that complement, not replace, the analysis of time-dependent systems. The objective of each section will be to simultaneously introduce some of the most important concepts in the study of dynamic systems and show how they can be improved upon using data-driven methods. A basic familiarity with dynamical systems is necessary to fully appreciate and understand the methods. Furthermore, since this book is united around illustrating how many data-driven methods can be used to understand, interpret, and produce new dynamical systems, it should not be considered as a comprehensive introduction to data science methods in general. This means that classical problems such as classification, image recognition, and independent component analysis will be absent from our discussion herein. The goal of this textbook is to demonstrate how innovative methods address gaps between theory and the understanding of specific models through interrogations that are inspired by dynamical systems theory.

The data that is produced to exemplify the methods throughout will be generated according to explicitly stated systems. Examples will focus on some of the most well-studied dynamical systems in the literature, which means we will have access to exact solutions and answers to our questions to gauge the performance of the methods. It should be noted that these example systems are only used to generate data, while methods are agnostic of such an underlying rule unless it is required for implementation. In summary, the methods we will encounter throughout this book are truly data driven in that they are seeded with data from a system but not necessarily information about the system that generates it. Thus, one may readily apply these methods to any dynamic data with the examples throughout this book acting as proofs-of-concept.

The presentation of this book emphasizes the mathematics behind the methods and demonstrates how they can be implemented numerically at only the highest level. In doing so, one will not find a full computational and algorithmic treatment of the numerical methods that underlie much of the implementation of the techniques. This is in part because topics such as numerical linear algebra and training neural networks already have extensive textbooks, courses, and resources to better understand them. Furthermore, the computational methods that we will rely upon in this book have been sufficiently developed by experts in both mathematics and computation to the point where implementing them can be done with premade functions in, for example, MATLAB or various Python packages (NumPy, SciPy, TensorFlow, etc.). Examples of this include computing the pseudoinverse in MATLAB with the `pinv()` function and training a neural network with stochastic gradient descent in TensorFlow. Thus, knowing what is needed for the specific task and having the correct software to implement it is sufficient for our demonstrations herein. The inclusion of a full code repository for these methods means that the presentation within this textbook will be primarily limited to the mathematical theory, while the use of premade functions can be found in the associated scripts and notebooks; see Appendix A. The reader is encouraged to follow along with the examples by running the associated code themselves and then repurposing the code to implement on their own systems and/or data.

Finally, the presented theory and methods will primarily be applied to discrete-time dynamical systems. This is for the simple reason that real-world data comes at discrete time intervals, no matter how fast the sampling rate. This means that even when continuous-time models are used for demonstration, we will only be sampling their state variables at discrete points in time, thus resulting in discrete-time data.

Chapter 2

Linear Evolution Models

In this chapter we will describe methods of obtaining linear models of evolution from data gathered from nonlinear systems. These methods are broadly organized around dynamic mode decomposition, which is presented in the first section of this chapter. As we progress through the sections of this chapter we will be introduced to more of the mathematical theory that underlies obtaining these linear models, primarily in the form of the Koopman operator. What one should be conscious of while moving through this chapter is that although the mathematical justification for the methods may be complex, little more is required to implement them computationally beyond matrix algebra. This means that these methods are often computationally cheap and efficient to implement, even on high-dimensional data.

Beyond the computational ease of implementing many of these methods, we are reminded of the benefits of a linear model. Unlike their nonlinear counterparts, the dynamics of linear systems can be entirely determined analytically. The study of linear systems completely reduces to identifying eigenvalues and eigenvectors to fully decompose the system into invariant subspaces. Moreover, one is only able to observe sinusoidal oscillations or exponential growth/decay in these invariant subspaces. Thus, finding linear descriptions of dynamic data is optimal, as the spectrum of the corresponding linear operator can be used to predict the long-time system dynamics.

2.1 ▪ Dynamic Mode Decomposition

We begin with Dynamic Mode Decomposition (DMD). The method aims to take advantage of low dimensionality in experimental data without having to rely on a given set of governing equations. DMD is primarily used to forecast data in time in a computationally cheap and efficient manner, progressing from one time to the next simply via matrix multiplication. That is, moving forward in time is achieved via a linear transformation, even if the underlying dynamics are nonlinear. In fact, it was for high-dimensional complex fluid flow models that Peter Schmid first proposed the method in 2008 [247]. Schmid was able to determine a linear transformation that maps successive instances of the fluid flow forward in time which faithfully reconstructs the expensive-to-generate synthetic data. Since Schmid's pioneering work there have been many improvements on the original DMD [248], but let us start with the method in its simplest form.

We begin with data matrices $X, Y \in \mathbb{C}^{M\times N}$, whose columns constitute *snapshots* through time. The snapshots are vectors in $\mathbb{C}^M$, which in the context of spatiotemporal systems are typically measurements of the system at different points in space. However, these snapshots could also simply represent measurements of different variables in the system, and so we make

Algorithm 2.1 Dynamic Mode Decomposition (DMD)

1. For each snapshot X_n, gather the subsequent snapshot Y_n
2. Arrange snapshots into data matrices $X = [X_1 \dots X_N]$ and $Y = [Y_1 \dots Y_N]$
3. Compute the DMD matrix $A = YX^\dagger$
4. Perform eigendecomposition $A\Psi = \mathcal{M}\Psi$
5. Construct mode decomposition (2.6)

no assumption of structure in them. Writing

$$X = \begin{bmatrix} | & | & \cdots & | \\ X_1 & X_2 & \cdots & X_N \\ | & | & \cdots & | \end{bmatrix}, \qquad Y = \begin{bmatrix} | & | & \cdots & | \\ Y_1 & Y_2 & \cdots & Y_N \\ | & | & \cdots & | \end{bmatrix}, \tag{2.1}$$

with each $X_n, Y_n \in \mathbb{C}^M$ being a snapshot, the critical assumption of DMD is that Y_n is the state exactly $h > 0$ time units in the future from X_n for each $1 \le n \le N$. One possibility is that we are collecting data from a single trajectory, wherein $Y_n = X_{n+1}$ for all n, but it is also possible to have snapshots of multiple trajectories represented in the matrices X and Y.

DMD seeks to find a matrix $A \in \mathbb{C}^{M\times M}$ that best maps the snapshots X_n to Y_n. That is, we seek A so that

$$Y_n \approx AX_n \tag{2.2}$$

for all $1 \le n \le N$. Mathematically (2.2) means that we are looking to identify a matrix A such that $\|Y_n - AX_n\|_2^2$ is as small as possible for all $1 \le n \le N$. Using the original data matrices we can state this as the minimization problem

$$\min_{A\in\mathbb{C}^{M\times M}} \|Y - AX\|_F^2, \tag{2.3}$$

where $\|\cdot\|_F$ is the Frobenius norm: the elementwise 2-norm of a matrix. The exact solution of the minimization problem (2.3) is given by

$$A = YX^\dagger = YV\Sigma^\dagger U^*, \tag{2.4}$$

where $\dagger$ indicates the pseudoinverse (or Moore–Penrose inverse) [100] and $X = U\Sigma V^*$ is the SVD of X. See Appendix B for details.

The DMD matrix $A = YX^\dagger$ that minimizes (2.3) provides a linear dynamical system,

$$Z_{n+1} = AZ_n, \tag{2.5}$$

for any $Z_n \in \mathbb{C}^M$ so that $AX_n \approx Y_n$. The reader should not confuse the above DMD process with linearization techniques used to understand system dynamics localized to a region of phase space. DMD is an entirely different process than linearization techniques that come from Taylor expansions and requires no assumption on the data in X and Y remaining in a small region of phase space. The result of DMD is a linear iterative scheme (2.5), and so its dynamics can be entirely understood through the eigenvalues and eigenvectors of A, similar to local linearization methods for dynamical systems.

Assuming for simplicity that A is diagonalizable, the M eigenvalue-eigenvector pairs $(\mu_m, \psi_m) \in \mathbb{C} \times \mathbb{C}^M$ allow for general iterations of (2.5) to be written as

$$Z_n = \sum_{m=1}^{M} c_m \mu_m^n \psi_m \tag{2.6}$$

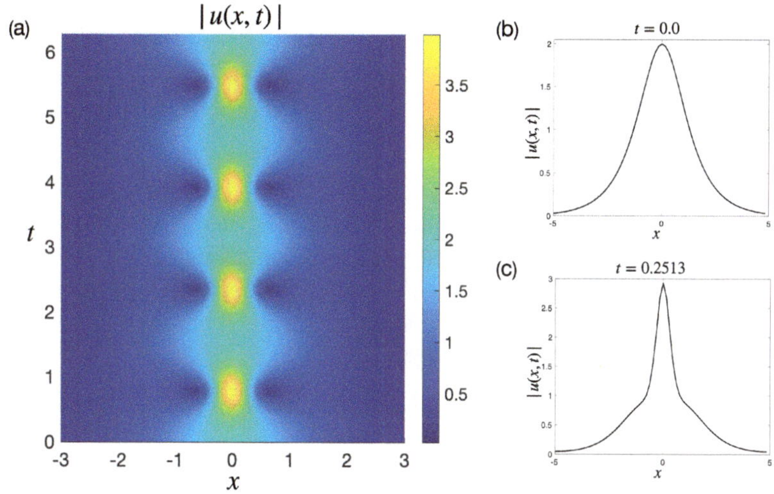

Figure 2.1. (a) *A space-time contour plot of* $|u(x,t)|$ *as a solution to the NLS* (2.8). (b) *The initial condition* $u(x,0) = 2\text{sech}(x)$. (c) *An instance of the NLS solution* $|u(x,t)|$ *at* $t = 0.2513$.

for some coefficients $c_m \in \mathbb{C}$. Much like POD, DMD provides a space-time separation of variables for the data. The eigenvectors ψ_m are referred to as the **DMD modes** and constitute macroscopic spatial patterns in the data. For example, when DMD is applied to a video file, one finds that eigenvectors with eigenvalues near 1 correspond to the background static motion in the video [107]. Generally, the eigenvalues determine the long-time fate of the DMD modes since

$$\lim_{n\to\infty} |\mu_m^n| = \begin{cases} 0, & |\mu_m| < 1, \\ 1, & |\mu_m| = 1, \\ \infty, & |\mu_m| > 1. \end{cases} \tag{2.7}$$

Long-time stable evolution under the DMD matrix A is captured by the eigenvalues on and inside the unit circle in the complex plane, while those outside of the unit circle lead to diverging iterates of the DMD model. The full DMD algorithm is provided in Algorithm 2.1.

Example 2.1. Let us follow as in [160] and use DMD to approximate the nonlinear Schrödinger equation (NLS)

$$iu_t + \frac{1}{2}u_{xx} + |u|^2 u = 0, \tag{2.8}$$

with $u(x,t)$ being a complex-valued function of space and time. The spatial domain will be $x \in [-5,5]$ with periodic boundary conditions on u. We seed DMD with data coming from a single trajectory with initial condition $u(x,0) = 2\text{sech}(x)$ integrated from $t = 0$ up to $t = 2\pi$. A space-time contour plot of $|u(x,t)|$ for this solution to the NLS (2.8) is presented in Figure 2.1 along with a plot of the initial condition and the profile $|u(x,t)|$ at $t = 0.2513$ for illustration. The resulting temporal dynamics are periodic with approximately four full periods taking place in the time window $t \in [0, 2\pi]$.

The data matrices X and Y are given by

$$X = \begin{bmatrix} u(x_1,t_1) & u(x_1,t_2) & \dots & u(x_1,t_N) \\ u(x_2,t_1) & u(x_2,t_2) & \dots & u(x_2,t_N) \\ \vdots & \vdots & \ddots & \vdots \\ u(x_M,t_1) & u(x_M,t_2) & \dots & u(x_M,t_N) \end{bmatrix} \tag{2.9}$$

and

$$Y = \begin{bmatrix} u(x_1, t_2) & u(x_1, t_3) & \dots & u(x_1, t_{N+1}) \\ u(x_2, t_2) & u(x_2, t_3) & \dots & u(x_2, t_{N+1}) \\ \vdots & \vdots & \ddots & \vdots \\ u(x_M, t_2) & u(x_M, t_3) & \dots & u(x_M, t_{N+1}) \end{bmatrix}, \tag{2.10}$$

where $x_1 < x_2 < \cdots < x_M$ are evenly spaced points in $[-5, 5]$ and $t_{n+1} = t_n + h$ for each $n = 1, \dots, N$. The experiments that follow will use $M = 64$ and $N = 20$, with $h = 2\pi/N = \pi/10$ and $t_1 = 0$, although results are similar for differing values of M and N.

The eigenvalues of the DMD matrix $A = YX^\dagger$ are plotted in Figure 2.2. There are $M = 64$ eigenvalues in total, with only $N = 20$ of them (approximately) nonzero. This is the maximal number of nonzero eigenvalues that A can have since

$$\operatorname{rank}(A = YX^\dagger) \leq \min\{\operatorname{rank}(Y), \operatorname{rank}(X^\dagger)\} = 20, \tag{2.11}$$

since both $X^\dagger$ and Y are rank 20 matrices. We check the accuracy of our DMD model using the iteration scheme (2.5) initiated with $Z_0 = X_1$ and compare $Z_N = A^N Z_0$ to the value of Y_N, the final snapshot in the training data. With our choices of $M = 64$ and $N = 20$ we get

$$\|Y_{20} - Z_{20}\|_2 = \texttt{1.2278} \times 10^{-9}, \tag{2.12}$$

an extremely accurate reconstruction of the nonlinear temporal evolution of (2.8) using only matrix multiplication. However, one can see from Figure 2.2 that there are eigenvalues of the DMD matrix outside of the unit circle, which means that sequences Z_n generated by (2.5) are expected to diverge as $n \to \infty$. Indeed, attempting to forecast beyond the training data to $t = 10\pi$ gives $\|Z_{100}\|_2 \geq \texttt{228}$, while to $t = 20\pi$ it gives $\|Z_{200}\|_2 \geq \texttt{80000}$. Since the snapshots remain uniformly bounded for all time, we see that our DMD forecast quickly becomes inaccurate because some eigenvalues of the DMD matrix lie outside the unit circle.

We can improve the long-time forecasting of our solution by incorporating some known physics of the NLS into our DMD model. Precisely, if $u(x, t)$ is a solution of (2.8) with periodic boundary conditions, integrating by parts gives

$$\begin{aligned} \frac{\mathrm{d}}{\mathrm{d}t} \int_{-5}^{5} |u(x,t)|^2 \mathrm{d}x &= \int_{-5}^{5} [u_t(x,t)\bar{u}(x,t) + u(x,t)\bar{u}_t(x,t)] \mathrm{d}x \\ &= \int_{-5}^{5} \left[\left(\frac{\mathrm{i}}{2} u_{xx}(x,t) + \mathrm{i}|u(x,t)|^2 u(x,t) \right) \bar{u}(x,t) \mathrm{d}x \right. \\ &\quad \left. + \int_{-5}^{5} u(x,t) \left(-\frac{\mathrm{i}}{2} \bar{u}_{xx}(x,t) - \mathrm{i}|u(x,t)|^2 \bar{u}(x,t) \right) \right] \mathrm{d}x \\ &= \frac{\mathrm{i}}{2} \int_{-5}^{5} [u_{xx}(x,t)\bar{u}(x,t) - \bar{u}_{xx}(x,t)u(x,t)] \mathrm{d}x \\ &= \underbrace{\frac{\mathrm{i}}{2} \Big[u_x(x,t)\bar{u}(x,t) - \bar{u}_x(x,t)u(x,t) \Big]_{x=-5}^{x=5}}_{=0} \\ &\quad - \frac{\mathrm{i}}{2} \int_{-5}^{5} \underbrace{\left[|u_x(x,t)|^2 - |u_x(x,t)|^2 \right]}_{=0} \mathrm{d}x \\ &= 0. \end{aligned}$$

This means that solutions of the NLS have $\int_{-5}^{5} |u(x,t)|^2 \mathrm{d}x$ being constant in time. In quantum mechanics, where the NLS comes from, this quantity is the *number of particles* in the

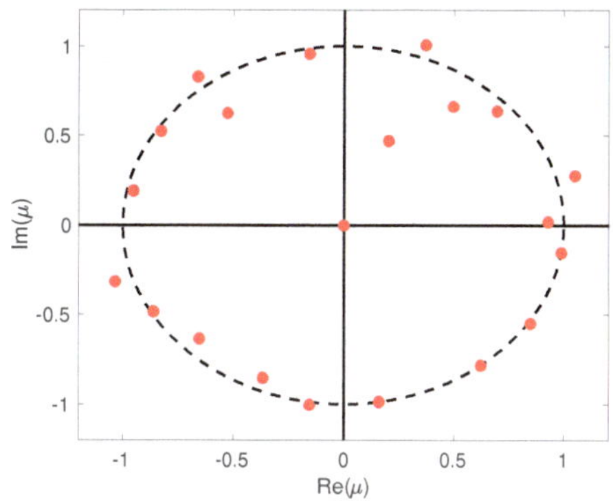

Figure 2.2. *Eigenvalues $\mu \in \mathbb{C}$ (red dots) of the DMD matrix resulting from data snapshots of solutions to the NLS* (2.8). *The unit circle is denoted by a dashed curve.*

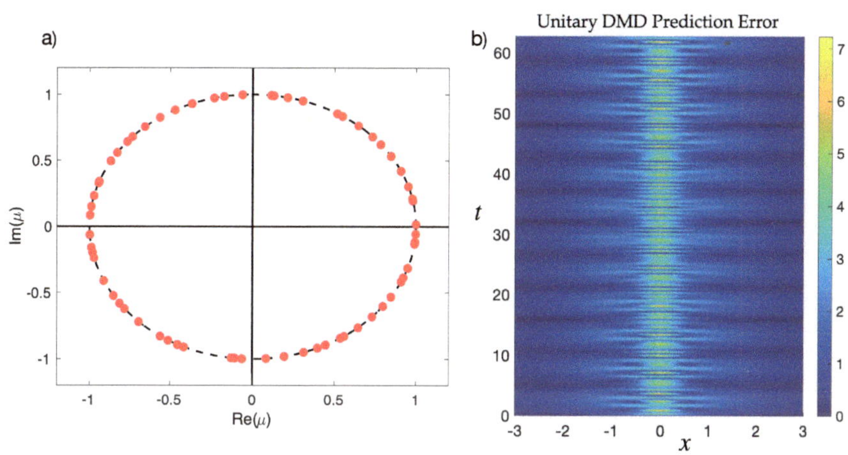

Figure 2.3. (a) *Eigenvalues $\mu \in \mathbb{C}$ (red dots) of the unitary DMD matrix solutions of the Procrustes problem* (2.13). (b) *Space-time contour plot of the prediction error $|u(x,t) - u_{\text{DMD}}(x,t)|$ for $u(x,t)$, a solution to the NLS* (2.8), *and $u_{DMD}(x,t)$, the unitary DMD matrix approximation.*

quantum system, and so our calculation above shows that this number remains the same over time. We can incorporate this property into our DMD forecast by enforcing the constraint that $\|AZ_n\|_2 = \|Z_n\|$ for all $n \geq 0$ since the norm $\|\cdot\|_2$ is the discrete analogue of the integral $\int_{-5}^{5} |u(x,t)|^2 \mathrm{d}x$. Such a constraint is equivalent to having our DMD matrix be *unitary*, meaning that $AA^* = A^*A = I$, where I is the identity matrix and $*$ denotes the Hermitian adjoint (conjugate transpose). The DMD minimization process (2.3) is now constrained by this unitary assumption, taking the form

$$\underset{AA^*=I}{\arg\min} \|Y - AX\|_F^2, \tag{2.13}$$

which reduces the search space for $M \times M$ DMD matrices.

The minimization problem (2.13) is an example of a **Procrustes problem**. Schönemann [250] proved that the solution of (2.13) is

$$A = U_{YX} V_{YX}^*, \tag{2.14}$$

where $U_{YX}\Sigma_{YX}V_{YX}^* = YX^*$ is the SVD. The eigenvalues of the unitary DMD matrix solution of (2.13) are plotted in Figure 2.3(a). Since all eigenvalues lie on the unit circle, there is no blowup or decay of the DMD modes, meaning that solutions to the DMD prediction scheme (2.5) remain bounded for all time. Figure 2.3(b) presents the error between the unitary DMD forecast with initial condition $Z_0 = X_1$, resulting in the approximation $u_{\text{DMD}}(x, t_n) = Z_n$, and the true time-periodic NLS soliton solution $u(x, t)$. Recall that the training data only goes up to $t = 2\pi$, while the forecast in Figure 2.3 goes far beyond this to $t = 20\pi$. The error remains bounded for all t, but the solution deviates significantly from the truth at times. Since (2.13) restricts the search space of matrices A, we have traded accurate short-time forecasts that blow up for slightly inaccurate long-time forecasts that remain bounded.

The previous example shows that incorporating physics into the DMD model can help forecast beyond the short training data time horizon. Furthermore, constraining the DMD matrix can potentially produce more accurate approximations of the snapshots, provide greater robustness to noise, and help generalize outside of the training data. Beyond just unitary matrices, Procrustes problems can incorporate known sparsity patterns, locality, causal dynamics, and symmetries, resulting in Physics-Informed DMD (piDMD) [22]. Beyond incorporating physics, the SVD can be used to produce low-rank approximations of high-dimensional ($M \gg 1$) datasets. This is achieved by keeping the first $r \ll M$ singular values and corresponding left and right singular vectors to produce an $r \times r$ DMD matrix instead of one that is much larger at $M \times M$ [111]. With so many extensions and the ease of applying many of them to data, DMD is a powerful tool to produce simple linear models from data to forecast, interpret, and control systems. However, DMD works best when M is large since the DMD matrix is large enough to capture nonlinear dynamics using a superposition of linearly evolving DMD modes. In Section 2.3, which follows, we will explore what happens when M is not large enough to obtain good DMD approximations and show how to inflate the value M to overcome this difficulty.

2.2 ▪ Sliding Window DMD

Many physical processes evolve on multiple timescales, often with simple and predictable evolution taking place over short periods of time and more complex drifting emerging over long periods. For example, the tides go in and out twice a day due to the moon's gravity, but small gravitational influences from the sun and other planets cause a slow drift in when the high and low tides arise that can only be observed over months or even years [289, 290]. We see a similar phenomenon of nearly periodic dynamics on short timescales with a complex drifting coming over long timescales throughout the physical sciences. Other examples include simulations of molecular dynamics [142, 14], atmospheric dynamics [212], and the motion of planets [46, 167, 186].

Multiscale dynamics present a difficulty for DMD. In typical applications DMD can identify the fast timescale motion, while the slow-scale drifting is difficult to realize with the standard DMD algorithm. Another setback when applying DMD to multiscale signals is that we require enough data to see the slow-scale motion emerge. This means that the number of snapshots N must be large, leading to a computational bottleneck when identifying the DMD matrix (2.4) using the pseudoinverse. Let us demonstrate some of the shortcomings with the following example.

Example 2.2. Following [77], we generate synthetic multiscale data from the system

$$\begin{aligned}
\dot{v}_1 &= v_2, \\
\dot{v}_2 &= -v_3^2 v_1^3, \\
\dot{v}_3 &= v_4, \\
\dot{v}_4 &= -100v_3 - 4v_3^3,
\end{aligned} \tag{2.15}$$

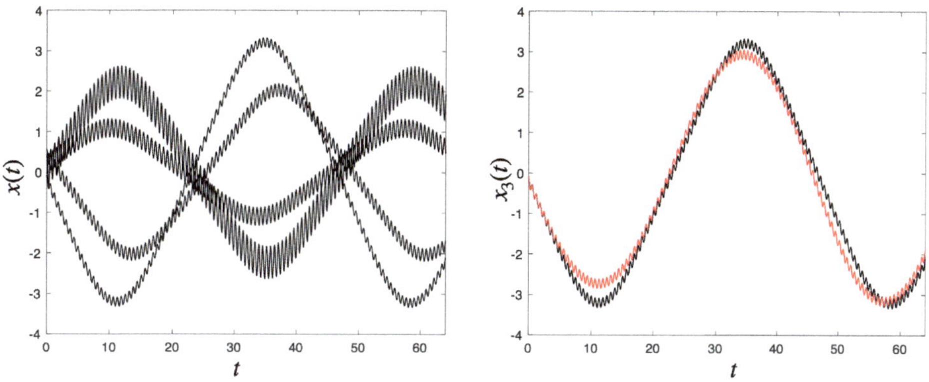

Figure 2.4. *Applying standard DMD to multiscale signals often leads to inaccurate predictions. Typically the fast-scale oscillations are identified, while the slower-scale drifts are inaccurate, resulting in a DMD forecast of the data that loses accuracy over time. Left: The four components of the mixed multiscale signal generated from* (2.15) *used as training data for standard DMD. Right: The DMD prediction (red) of the* $x_3(t)$ *component (black) of the multiscale data.*

with initial condition $(v_1(0), v_2(0), v_3(0), v_4(0)) = (0, 0.5, 0, 0.5)$. The variables (v_3, v_4) decouple to form a weakly nonlinear undamped Duffing/cubic oscillator. These variables constitute the fast-scale dynamics of the system, oscillating with a period near 0.1. Their oscillation is coupled to the (v_1, v_2) components to build the slow timescale dynamics, again in the form of a cubic oscillator. To thoroughly mix the fast and slow dynamics of system (2.15) we generate a random orthogonal matrix $Q \in \mathbb{R}^{4\times4}$ to construct our multiscale data:

$$\begin{bmatrix} x_1(t) \\ x_2(t) \\ x_3(t) \\ x_4(t) \end{bmatrix} = Q \begin{bmatrix} v_1(t) \\ v_2(t) \\ v_3(t) \\ v_4(t) \end{bmatrix}. \tag{2.16}$$

Training data is obtained integrating (2.15) with a timestep of $h = 10^{-4}$ from $t = 0$ to $t = 64$. The DMD matrices X and Y in (2.1) have $M = 4$ and $N = 64000$ and $Y_n = X_{n+1}$ for each $n = 1, \ldots, N$. The training data is plotted in the left panel of Figure 2.4.

Applying DMD to this synthetic multiscale data results in the DMD matrix

$$A = \begin{bmatrix} \texttt{0.99961} & \texttt{0.00101} & \texttt{0.00173} & \texttt{0.002383} \\ -\texttt{0.00089} & \texttt{1.00255} & \texttt{0.00437} & \texttt{0.00621} \\ -\texttt{0.00033} & \texttt{0.00067} & \texttt{1.00123} & \texttt{0.001707} \\ \texttt{0.00051} & -\texttt{0.00150} & -\texttt{0.00247} & \texttt{0.99661} \end{bmatrix}. \tag{2.17}$$

The resemblance of the DMD matrix to the identity comes from the fact that we are taking small steps between successive snapshots, and therefore little is changing from one step to the next. In the right panel of Figure 2.4 we provide the DMD forecast against the training data for $x_3(t)$, with results for the other components being similar. While the fast-scale motion is captured by DMD, the slow background evolution is inaccurate and becomes worse as time goes on.

The previous example illustrates some of the failure that should be expected when applying DMD to multiscale data. To overcome this, we consider **sliding window DMD** [77]. This method draws inspiration from short-time Fourier transform methods, such as the Gabor transform (see [160] and the references therein). However, it leverages the main advantage of using

Algorithm 2.2 Sliding Window DMD

1. Gather sequential snapshot data $X_1, X_2, \dots, X_N$
2. Fix window size $T \geq 1$ and construct matrices $X^{(k)} = [X_k \dots X_{k+T}]$
3. For each k:
   (a) Compute the DMD matrix for each window $A^{(k)} = X^{(k+1)}[X^{(k)}]^\dagger$
   (b) Eigendecomposition to get DMD modes $A^{(k)}\Psi^{(k)} = \mathcal{M}^{(k)}\Psi^{(k)}$
   (c) Define threshold $\lambda > 0$ and separate fast and slow modes according to (2.21)
   (d) Construct $x^{(k)}_{\text{fast}}(t)$ and $x^{(k)}_{\text{slow}}(t)$ from separated DMD modes
4. Apply clustering over k of DMD eigenvalues to find dominant timescales
5. Construct separated global fast and slow components using each $x^{(k)}_{\text{fast}}(t)$ and $x^{(k)}_{\text{slow}}(t)$

DMD: It can account for both temporal and spatial/variable coherence, as identified by the eigenvalues and eigenvectors of the DMD matrix, respectively.

To begin, we assume we have samples from a continuous signal $x(t) \in \mathbb{R}^M$ with a sampling rate of $h > 0$. To seed DMD we take snapshots $X_n := x(h(n-1))$ for each $n = 1, \dots, N+1$. Then we divide the data up into "windows" of size $T \geq 1$ so that for each $k = 1, \dots, N-T$ we have

$$X^{(k)} = \begin{bmatrix} | & | & & | \\ X_k & X_{k+1} & \cdots & X_{k+T} \\ | & | & & | \end{bmatrix} \in \mathbb{C}^{M \times T}; \tag{2.18}$$

i.e., the columns of $X^{(k)}$ are the kth through $(k+T)$th snapshots. An ideal window length would be long enough to capture some variations in the fastest timescale but short enough to not fully resolve the slow timescale. For each k we can DMD to identify a matrix $A^{(k)} := X^{(k+1)}[X^{(k)}]^\dagger$ so that $X_n^{(k+1)} \approx A^{(k)}X_n^{(k)}$. Each DMD matrix $A^{(k)}$ has M eigenvalues associated to it, and due to the eigendecomposition of the matrix we can recover the signal in the window by

$$x^{(k)}(t) = \sum_{m=1}^{M} c_m^{(k)} \psi_m^{(k)} e^{\omega_m^{(k)} t}, \tag{2.19}$$

following (2.6). Here $\omega_m^{(k)} := \ln(\mu_m^{(k)})/h$ from the eigenvalues $\mu_m^{(k)}$ of the DMD modes $\psi_m^{(k)}$.

With DMD models of the signal over each window, the spectra of the DMD matrices $A^{(k)}$ are then clustered to discover the dominant frequency content. Precisely, for each $k = 1, \dots, N - T$ we concatenate the $|\omega_m^{(k)}|^2$ into a vector and apply a clustering algorithm. The reason for clustering $|\omega_m^{(k)}|^2$ instead of $|\omega_m^{(k)}|$ is that the exponent helps to further separate our the fast and slow frequencies, thus making clustering easier [77]. We will see in Section 3.4 that the ability to clearly identify the fast period of oscillation in multiscale signals can be particularly useful in identifying equations of motion that govern the slow timescale drift.

The DMD solutions in each window (2.19) can be used to reconstruct the full signal with

$$x_{\text{global}}(t) = \frac{\sum_{k=1}^{N-T} e^{-(t-\eta_k)^2/\sigma^2} x^{(k)}(t)}{\sum_{k=1}^{N-T} e^{-(t-\eta_k)^2/\sigma^2}}, \tag{2.20}$$

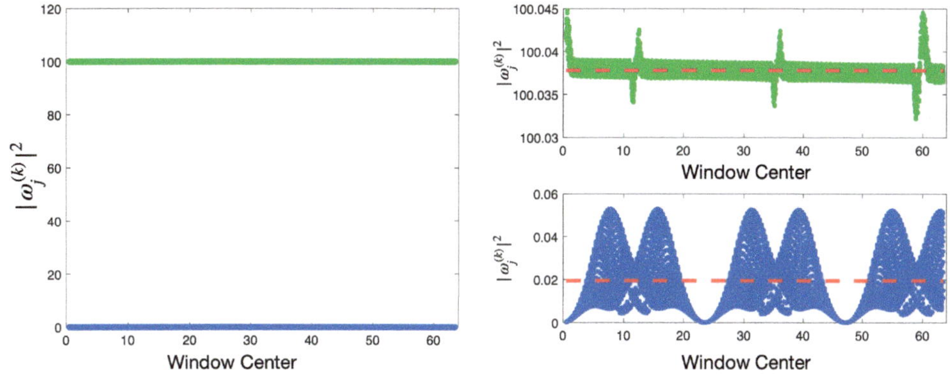

Figure 2.5. *The result of a clustering the square modulus of the continuous-time frequencies $\omega_j^{(k)}$ using sliding window DMD. There is a clear delineation between fast (green) and slow (blue) frequencies. The right panel isolates each frequency cluster and plots the cluster mean (red), identified with k-means.*

with $\eta_k := (k-1+T)/2$ representing the middle of each window. We have taken a weighted sum of the windowed reconstructions $x^{(k)}(t)$, where the weight is given by a Gaussian function which acts to smoothly localize each reconstruction in its appropriate window. Per [77], the standard deviation of the Gaussian is taken to be $1/8$th of the windows width in time, $\sigma = Th/8$.

Finally, the windowed reconstructions (2.19) can be used to separate out the fast- and slow-scale dynamics. Indeed, consider a threshold $\lambda > 0$, and define the set

$$\mathcal{I}_{\text{fast}}^{(k)} := \{m : |\omega_m^{(k)}| \geq \lambda\}. \tag{2.21}$$

The reconstructed fast timescale dynamics in window k are given by

$$x_{\text{fast}}^{(k)}(t) = \sum_{m \in \mathcal{I}_{\text{fast}}} c_m^{(k)} \psi_m^{(k)} \mathrm{e}^{\mathrm{i}\omega_m^{(k)} t}. \tag{2.22}$$

The linearity of the DMD eigenfunction expansion guarantees that the slow dynamics in window k are given by what remains of the signal after subtracting out the fast dynamics,

$$x_{\text{slow}}^{(k)}(t) = x^{(k)}(t) - x_{\text{fast}}^{(k)}(t). \tag{2.23}$$

If the slow dynamics themselves evolve on multiple timescales, one may further subdivide the slow frequencies using another threshold. Importantly, while λ is user-defined, the choice of it should not be complicated so long as the fast and slow timescales are clearly delineated. This process is summarized in Algorithm 2.2. We demonstrate with the following example.

Example 2.3. Let us return to the signal in Example 2.2 and apply sliding window DMD to it. We consider a window width of 1 time unit, meaning that $T = 1/h = 10^4$. To identify the frequency clusters we use k-means to identify two clusters: one fast and one slow. Results are plotted in Figure 2.5. Using larger windows means that the slow evolution has more of an influence within each window, potentially leading to the same inaccuracy we saw when applying DMD to the multiscale signal previously. Conversely, shorter windows risk being so small that less than a period of the fast oscillation is in it, likely leading DMD to fail to identify them. Therefore, there is a Goldilocks zone for the choice of the window size—not too small and not too big. Dylewsky et al. suggest a window size that sees at least two full periods of the fast oscillations [77] , which should be small enough for the slow drift to remain nearly constant.

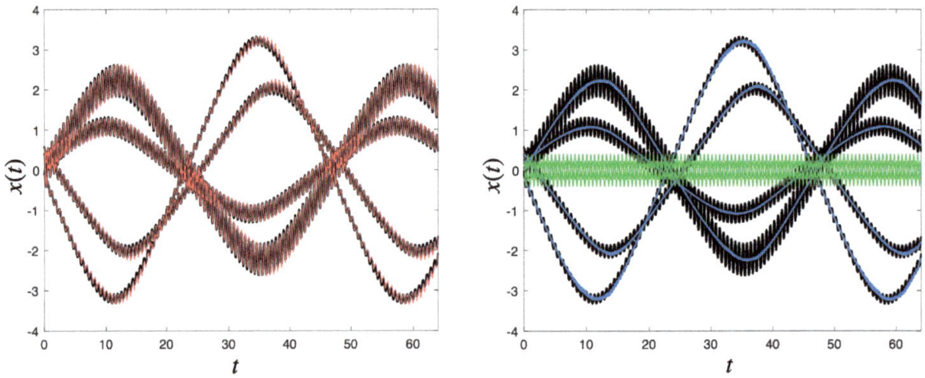

Figure 2.6. *Left: A reconstruction (red) of the multiscale signal (black) using sliding window DMD. Right: Sliding window DMD can be used to separate out the fast (green) from the slow (blue) evolution of the multiscale signal.*

Frequency cluster centroids are identified as `100.03783` and `0.01948`, plotted as a horizontal red line in the right panel of Figure 2.5. Thus, the fast scale has frequency $\omega_{\text{fast}} \approx$ `10.00189`, while the slow frequency is $\omega_{\text{slow}} \approx$ `0.13959` in the signal. For comparison, the DMD matrix (2.17) returns the same fast frequency (to the reported decimal places), while the slow frequency is found to be `0.13580`. Despite this being close to the sliding window method's slow frequency, the difference is noticeable as time goes on. This can be seen in Figure 2.4, where the slow drift quickly falls out of sync with training data. Compare this with Figure 2.6, where both the fast- and slow-scale motion closely track the original signal. Furthermore, in Figure 2.6 we demonstrate the scale separation afforded by the linearity of DMD using (2.22) and (2.23). This allows one to forecast the slow evolution of the signal without the noise of a rapidly oscillating fast component.

2.3 ▪ Time-Delay Coordinates

At the simplest conceptual level, the accuracy of DMD relies on having high-dimensional measurements. This allows for the discovered DMD matrix to have numerous different eigendirections which when put together can approximate highly nonlinear datasets. However, there are many scenarios in which only scalar or low-dimensional measurements of the system are available, potentially leading to a DMD model that has no hope of forecasting complex data. To illustrate, suppose we sample the signal $x(t) = \sin(t)$ at a rate $h > 0$. Our measurements are scalar $x_n := x(nh)$, $n = 0, \dots, N$, leading to a DMD matrix that is a single real variable:

$$x_{n+1} = ax_n, \qquad a \in \mathbb{R}. \tag{2.24}$$

The general term of (2.24) is $x_n = a^n x_0$, which exhibits exponential growth or decay but never the desired oscillations. Therefore, applying DMD to the snapshots $x_n = \sin(nh)$ is useless.

Although the above example is not meant to be realistic, it illustrates something very important when it comes to applying DMD to real-world data. What if we have incomplete information about a system? That is, there may be a number of variables associated to our system of interest, but only a few can be measured. A related issue would be for spatiotemporal systems wherein we can only measure the system at a limited number of points in space but not nearly enough to obtain a complete picture of the dynamics across space. This problem has similarly arisen in the reconstruction of chaotic attractors from partial measurements, with Takens showing that faithful reconstruction can be achieved using **time-delay coordinate embeddings** [269]. In this section we leverage Takens's ideas and demonstrate how delay coordinate embeddings are a useful method of inflating the dimension of observations to produce more accurate DMD models. Beyond the applications in this section, these methods have been successfully used to apply DMD to data from power systems [11], epileptic seizures [267], traffic [20], and the weather [281]. Time-delay coordinates can also be used for control [76, 16], while the performance of DMD on time-delay coordinates has been compared to other DMD methods in [268, 81].

To begin, suppose we are given a set of measurements $x_n \in \mathbb{C}^d$, $n = 1, \ldots N$. The work [213] has proposed fitting $A_j \in \mathbb{R}^{d\times d}$, $j = 1, \ldots, \tau$, to identify a linear delay system

$$x_{n+1} = A_0 x_n + A_1 x_{n-1} + \cdots + A_\tau x_{n-\tau} \tag{2.25}$$

for some $\tau \geq 1$. The process (2.25) can alternatively be stated as a constrained version of the usual DMD minimization problem (2.3) by arranging the data into **Hankel** matrices:

$$X = \begin{bmatrix} x_1 & x_2 & \cdots & x_{N-\tau-1} \\ x_2 & x_3 & \cdots & x_{N-\tau} \\ \vdots & \vdots & \ddots & \vdots \\ x_\tau & x_{\tau+1} & \cdots & x_{N-1} \end{bmatrix}, \quad Y = \begin{bmatrix} x_2 & x_3 & \cdots & x_{N-\tau} \\ x_3 & x_4 & \cdots & x_{N-\tau+1} \\ \vdots & \vdots & \ddots & \vdots \\ x_{\tau+1} & x_{\tau+2} & \cdots & x_N \end{bmatrix} \in \mathbb{C}^{d\tau\times N-\tau-1}. \tag{2.26}$$

Notice that the rows and columns of X and Y are successive snapshots of the data. Then we constrain the DMD matrix to take the form

$$A = \begin{bmatrix} 0_d & I_d & 0_d & \cdots & 0_d & 0_d \\ 0_d & 0_d & I_d & \cdots & 0_d & 0_d \\ \vdots & \vdots & \vdots & \ddots & \vdots & \vdots \\ 0_d & 0_d & 0_d & \cdots & 0_d & I_d \\ A_\tau & A_{\tau-1} & A_{\tau-2} & \cdots & A_1 & A_0 \end{bmatrix}, \tag{2.27}$$

where 0_d is the $d \times d$ zero matrix and I_d is the $d \times d$ identity matrix. The final row of A describes the linear delay system (2.25), while the previous rows simply push the delayed data forward in time. The form (2.25) has long been used in the timeseries analysis community for autoregressive and moving average (ARMA) models [37]. Importantly, [213, Theorem 1] details that any periodic, uniformly sampled, scalar signal x_n can be perfectly recovered using τ equal to one less than the number of nonzero coefficients in the Fourier spectrum of the signal.

Example 2.4. Let us return to the example $x_n = \sin(nh)$ from above and notice that

$$\sin(nh) = \frac{1}{2\mathrm{i}}\left(\mathrm{e}^{\mathrm{i}nh} - \mathrm{e}^{-\mathrm{i}nh}\right). \tag{2.28}$$

Taking $\tau = 1$, we are required to find $a_0, a_1 \in \mathbb{R}$ to fit

$$\sin((n+1)h) = a_0 \sin(nh) + a_1 \sin((n-1)h). \tag{2.29}$$

Using the identity $\sin(u + v) = \sin(u)\cos(v) + \cos(u)\sin(v)$ one finds that $a_0 = 2\cos(h)$ and $a_1 = -1$. The resulting time-delay coordinate DMD system (2.27) is given by

$$\begin{bmatrix} x_n \\ x_{n+1} \end{bmatrix} = \begin{bmatrix} 0 & 1 \\ -1 & 2\cos(h) \end{bmatrix} \begin{bmatrix} x_{n-1} \\ x_n \end{bmatrix}, \tag{2.30}$$

for which the eigenvalues of the DMD matrix are $\mathrm{e}^{\pm ih}$. The nth exponent of these eigenvalues appear in (2.28) for the signal n steps into the future.

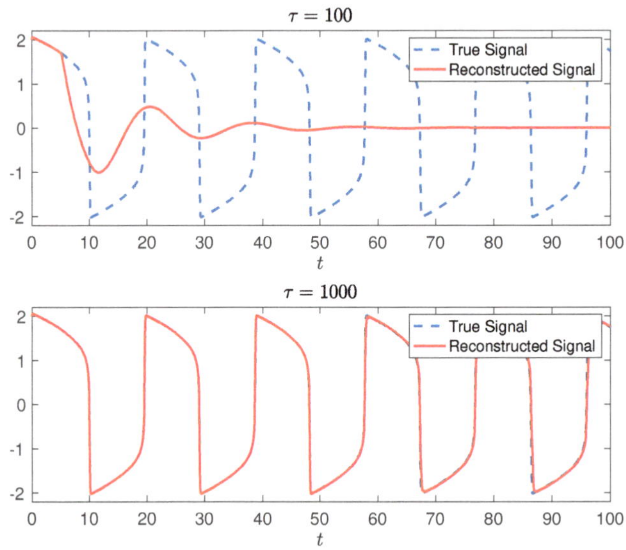

Figure 2.7. *Forecasting solutions of the Van der Pol oscillator* (2.31) *using delay-coordinate DMD* (2.25) *requires a large number of delays τ to obtain accurate predictions for long times. Larger values of τ can forecast the signal for longer, but ultimately finite, times.*

Example 2.5. Let us now examine a periodic signal that is not the superposition of Fourier modes. We consider partial observations of the second-order Van der Pol oscillator [280]

$$\ddot{x} + 10(x^2 - 1)\dot{x} + x = 0. \tag{2.31}$$

Equation (2.31) is known to fall into a steady periodic oscillation [109], and we will take our partial observations to be the state variable $x(t)$. As in the previous example, we do not have enough observed variables to produce an oscillating linear model with DMD, and so we use time-delay coordinates to inflate the dimension of our observations.

We take snapshots to be $x_n = x(nh)$ with a sampling rate $h = 0.05$ and $n = 0, \ldots, 4000$. The highly nonlinear oscillation is shown in Figure 2.7 and does not have a sparse Fourier spectrum. Therefore, the results of [213] cannot be applied to guarantee that we can perfectly reconstruct the signal. In Figure 2.7 we compare forecasts of the initial condition $(x(0), \dot{x}(0)) = (2, 2)$ using the discovered delay-coordinate DMD scheme (2.25) for $\tau = 100$ and $\tau = 1000$. With $\tau = 100$ the forecast quickly becomes inaccurate, and the prediction converges to zero. However, with $\tau = 1000$ we can forecast three full periods of oscillation before errors begin to become noticeable. No finite value of τ will be able to capture the signal for all time since $x(t)$ requires infinitely many Fourier terms to reconstruct, but larger and larger values of τ will allow one to forecast further and further into the future.

The SVD of the Hankel data matrix, $X = U\Sigma V^*$, can be used to produce *observables* for the data x_n. The columns of $U \in \mathbb{C}^{d\tau \times d\tau}$ constitute the new observables in the delay coordinates, and the columns of $V \in \mathbb{C}^{(N-\tau-1)\times(N-\tau-1)}$ describe their temporal evolution [111, 49]. Letting $U_r \in \mathbb{C}^{d\tau \times r}$ be the first r columns of U, we may define

$$\tilde{X} = U_r^* X, \qquad \tilde{Y} = U_r^* Y \tag{2.32}$$

to be the projections of X and Y, respectively, onto the first r observables. We may then apply DMD (2.3) to $\tilde{X}$ and $\tilde{Y}$ instead of X and Y. Since $X = U_r\tilde{X}$, one may always recover the

Algorithm 2.3 DMD with Time-Delayed Observables

1. Gather sequential snapshot data $x_1, x_2, \ldots, x_N$
2. Construct Hankel matrices X and Y in (2.26) from snapshots
3. Compute SVD $X = U\Sigma V^*$
4. Fix $r \geq 1$ and project onto first r left singular vectors: $\tilde{X} = U_r^* X$ and $\tilde{Y} = U_r^* Y$
5. Obtain DMD matrix $A = \tilde{Y}\tilde{X}^\dagger$
6. Perform eigendecomposition $A\Psi = \mathcal{M}\Psi$
7. Recover DMD modes corresponding to X and Y from $U_r\Psi$

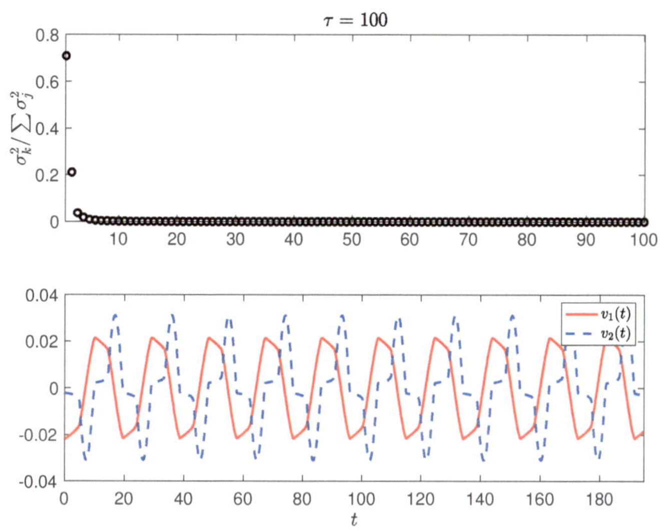

Figure 2.8. *Top: The "energy" of each singular value of the Hankel matrix* (2.26) *obtained from* $\tau = 100$ *delays of an observation of the Van der Pol oscillator* (2.31)*. Bottom: The temporal evolution of the first two observations obtained from the SVD of the Hankel matrix.*

original delayed variables of the Hankel matrix X by left multiplying DMD forecasts of $\tilde{X}$ by U_r. This process is summarized in Algorithm 2.3. The following example is meant to demonstrate why such a process is advantageous, while the following subsection goes into more detail regarding the observables of a state-space variable.

Example 2.6. Consider again the Van der Pol oscillator from Example 2.5 with the same snapshot matrices X and Y. In Figure 2.8 we present the energy of each singular value, given by the fraction

$$\frac{\sigma_k^2}{\sum_{j=1}^{\min\{M,N\}} \sigma_j^2}, \qquad k = 1, \ldots, \min\{M, N\}, \tag{2.33}$$

of the SVD of X using $\tau = 100$ delays. The resulting temporal dynamics of the leading two observables are also presented, which are again periodic although quite complicated. Hence, as was the case in Example 2.5 for the original data x_n, we should not expect DMD to be able to

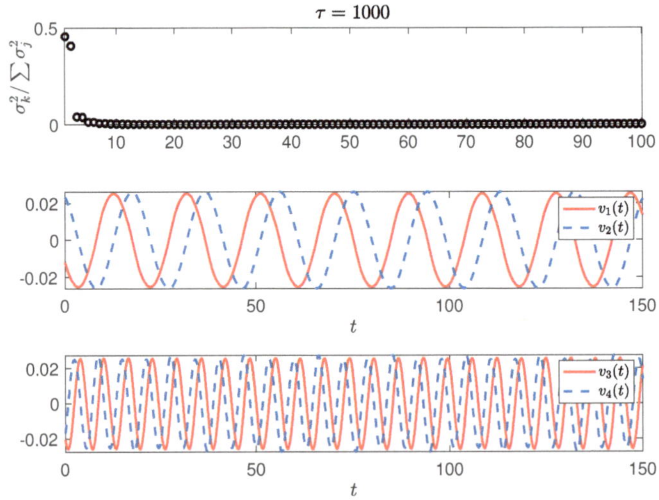

Figure 2.9. *Top: The "energy" ratio of each singular value of the Hankel matrix* (2.26) *obtained from* $\tau = 1000$ *delays of an observation of the Van der Pol oscillator* (2.31). *Middle and bottom: The temporal evolution of the first four observations obtained from the SVD of the Hankel matrix. By increasing the number of delays the observations now oscillate nearly linearly in time.*

forecast these nonlinear oscillations for all time. Increasing the number of delays to $\tau = 1000$ results in a singular value spectrum that approximately comes in pairs, with the corresponding temporal fluctuations of the observations being nearly linear oscillations. This is shown in Figure 2.9, and one can see that the third and fourth observations oscillate at a higher frequency than the first and second observations. Projecting the data matrices X and Y onto the columns of U_r with $r = 4$ results in temporal dynamics that are (approximately) linearly oscillating. In turn, we can nearly perfectly reconstruct the nonlinear periodic Van der Pol signal by first applying DMD to $\tilde{X}$ and $\tilde{Y}$, as defined in (2.32). Furthermore, the first four singular values constitute 94% of the energy, meaning little information about the Van der Pol dynamics (2.31) is lost when projecting onto only the first four dominant observables.

2.4 ▪ Approximating the Koopman Operator

In the previous subsection we saw that the SVD to a data matrix $X = U\Sigma V^*$ results in a set of observables in the columns of U that evolve in time according to the columns of V. Furthermore, we saw in Example 2.6 that by increasing the number of delays τ the dominant observables appear to exhibit linear oscillations, i.e., like sines and cosines. Therefore, while the dynamics are likely governed by a nonlinear process, finding the right observables can transform the data in such a way that it now evolves (approximately) linearly.

Building upon this, suppose that we collect data governed by the discrete dynamical system $F : \mathcal{M} \to \mathcal{M}$, with $\mathcal{M} \subseteq \mathbb{R}^M$. We again consider data matrices X and Y in (2.1) with the assumption that $Y_n = F(X_n)$ for all columns X_n of X and Y_n of Y. An **observable** is a scalar function on the state space, $\phi : \mathcal{M} \to \mathbb{C}$, which we apply to the data in X and Y, resulting in the observed data vectors

$$\Phi(X) = \begin{bmatrix}\phi(X_1) & \phi(X_2) & \cdots & \phi(X_N)\end{bmatrix} \in \mathbb{C}^{1\times N} \tag{2.34}$$

and

$$\Phi(Y) = \begin{bmatrix}\phi(Y_1) & \phi(Y_2) & \cdots & \phi(Y_N)\end{bmatrix} \in \mathbb{C}^{1\times N}. \tag{2.35}$$

Importantly, with the assumption that $X_n \mapsto Y_n$ by F, we have that $\phi(Y_n) = \phi(F(X_n))$ for all $1 \leq n \leq N$. Hence, all temporal evolution takes place inside of the observables via the function composition $\phi \circ F$.

It is important to note that for any observables $\phi, \psi : \mathcal{M} \to \mathbb{C}$ we have the linearity property

$$(\alpha\phi + \beta\psi) \circ F = \alpha\phi \circ F + \beta\psi \circ F, \qquad \forall \alpha, \beta \in \mathbb{C}. \tag{2.36}$$

Thus, complex observations can be built up as linear combinations of a base set. This process of advancing iterates of F inside of observables was first proposed by B. O. Koopman [150] and has resulted in the **Koopman operator**, which acts by

$$\mathcal{K}\phi = \phi \circ F \tag{2.37}$$

on observables ϕ belonging to a suitable space of functions. By (2.36), the Koopman operator represents a linear dynamical system, which like DMD can be fully described in terms of its spectrum. The trade-off here now is that since $\mathcal{K}$ acts on functions, it is infinite dimensional even though F is finite dimensional.

What may not be clear yet is what the Koopman operator is good for. Although F and $\mathcal{K}$ appear quite different, they are just two different representations of the same fundamental behavior. The link is made through the "state-space observables" $g_m(x) = x_m$ for each $x = (x_1, \ldots, x_M) \in \mathcal{M}$. Since $\mathcal{K}$ is linear, we suppose that $\{(\mu_\alpha, \phi_\alpha)\}_{\alpha \in A}$ is a possibly infinite collection of its eigenvalues and eigenfunctions. Assuming each g_m is in the span of the set of eigenfunctions,

$$x_m = g_m(x) = \sum_{\alpha \in A} c_{m,\alpha}\phi_\alpha(x), \qquad c_{m,\alpha} \in \mathbb{C}, \quad m = 1, \ldots, M, \tag{2.38}$$

in turn gives that the nonlinear function $F = (g_1 \circ F, \ldots, g_M \circ F)^\mathsf{T}$ is linearized by

$$(g_m \circ F)(x) = [\mathcal{K}g_m](x) = \sum_{\alpha \in A} \mu_\alpha c_{m,\alpha}\phi_\alpha(x), \qquad m = 1, \ldots, M, \tag{2.39}$$

using the eigenfunction property $\mathcal{K}\phi_\alpha = \mu_\alpha\phi_\alpha$ for each $\alpha \in A$. The coefficients $[c_{m,\alpha}]_{\alpha \in A}$ are referred to as the **Koopman modes**, which describe how the Koopman eigenfunctions come together to linearize the system. Bringing this back to DMD, (2.39) shows that nonlinear dynamics can be posed as linear evolution so long as one has access to the appropriate observables to transform the data. What follows is a simple analytical example adapted from [90, Section III].

Example 2.7. For any $\lambda_1, \lambda_2 \in \mathbb{R}$, consider the discrete dynamical system

$$\begin{bmatrix} x_1 \\ x_2 \end{bmatrix}_{n+1} = \begin{bmatrix} \lambda_1 x_1 \\ \lambda_2(x_2 - x_1^2) \end{bmatrix}_n. \tag{2.40}$$

While (2.40) is nonlinear, we will show that it only takes a finite number of Koopman eigenfunctions to linearize the system via (2.39). For general systems one should expect that the eigenfunction expansion (2.39) requires infinitely many nonzero c_α. However, in the case of (2.40) the associated Koopman operator has an invariant three-dimensional subspace for which the state-space observables $g_1(x_1, x_2) = x_1$ and $g_2(x_1, x_2) = x_2$ belong. The result is that we can recast the nonlinear dynamics of (2.40) as a higher but finite dimensional linear system.

If we consider the observables $y_1 = x_1$, $y_2 = x_2$, and $y_3 = x_1^2$, then (2.40) can be rewritten as an equivalent linear system,

$$\begin{bmatrix} y_1 \\ y_2 \\ y_3 \end{bmatrix}_{n+1} = \begin{bmatrix} \lambda_1 & 0 & 0 \\ 0 & \lambda_2 & -\lambda_2 \\ 0 & 0 & \lambda_1^2 \end{bmatrix} \begin{bmatrix} y_1 \\ y_2 \\ y_3 \end{bmatrix}_n. \tag{2.41}$$

This linear system represents the restriction of the Koopman operator to the invariant linear subspace spanned by the observables $\{x_1, x_2, x_1^2\}$. The matrix in (2.41) has eigenvalues λ_1, λ_2, and λ_1^2 with corresponding eigenvectors

$$\begin{aligned} v_{\lambda_1} &= \begin{bmatrix} 1 & 0 & 0 \end{bmatrix}, \\ v_{\lambda_2} &= \begin{bmatrix} 0 & 1 & 0 \end{bmatrix}, \\ v_{\lambda_1^2} &= \begin{bmatrix} 0 & \frac{\lambda_2}{\lambda_2 - \lambda_1^2} & 1 \end{bmatrix}. \end{aligned} \tag{2.42}$$

The corresponding left (adjoint) eigenvectors of the matrix in (2.41) are

$$\begin{aligned} w_{\lambda_1} &= \begin{bmatrix} 1 & 0 & 0 \end{bmatrix}, \\ w_{\lambda_2} &= \begin{bmatrix} 0 & 1 & \frac{-\lambda_2}{\lambda_2 - \lambda_1^2} \end{bmatrix}, \\ w_{\lambda_1^2} &= \begin{bmatrix} 0 & 0 & 1 \end{bmatrix}, \end{aligned} \tag{2.43}$$

which have been scaled so that $\langle w_i, v_j \rangle = \delta_{i,j}$. We can use these left eigenvectors to produce three eigenfunctions of the Koopman operator by

$$\begin{aligned} \psi_{\lambda_1}(x_1, x_2) &= \langle y, w_{\lambda_1} \rangle = y_1 = x_1, \\ \psi_{\lambda_2}(x_1, x_2) &= \langle y, w_{\lambda_2} \rangle = y_2 - \frac{\lambda_2}{\lambda_2 - \lambda_1^2} y_3 = x_2 - \frac{\lambda_2}{\lambda_2 - \lambda_1^2} x_1^2, \\ \psi_{\lambda_1^2}(x_1, x_2) &= \langle y, w_{\lambda_1^2} \rangle = y_3 = x_1^2, \end{aligned} \tag{2.44}$$

where $y = [y_1, y_2, y_3]^\intercal$.

These Koopman eigenfunctions completely diagonalize (2.41), giving

$$\begin{bmatrix} \psi_{\lambda_1} \\ \psi_{\lambda_2} \\ \psi_{\lambda_1^2} \end{bmatrix}_{n+1} = \begin{bmatrix} \lambda_1 & 0 & 0 \\ 0 & \lambda_2 & 0 \\ 0 & 0 & \lambda_1^2 \end{bmatrix} \begin{bmatrix} \psi_{\lambda_1} \\ \psi_{\lambda_2} \\ \psi_{\lambda_1^2} \end{bmatrix}_{n}. \tag{2.45}$$

The full nonlinear state dynamics (2.40) are recovered by

$$\begin{aligned} x_1 &= \psi_{\lambda_1}(x_1, x_2), \\ x_2 &= \psi_{\lambda_2}(x_1, x_2) + \left(\frac{\lambda_2}{\lambda_2 - \lambda_1^2} \right) \psi_{\lambda_1^2}(x_1, x_2). \end{aligned} \tag{2.46}$$

Thus, (2.46) provides an equivalent linear representation for the dynamics of (2.40) wherein iterations of (x_1, x_2) come from multiplying each eigenfunction by its corresponding eigenvalue.

Despite the ease of Example 2.7, for most systems we should expect to require infinitely many Koopman eigenfunctions to fully linearize the dynamics of a nonlinear system. While this is not something we can do in practice, **Extended Dynamic Mode Decomposition** (EDMD) leverages the Koopman operator framework to provide a data-driven method of approximating the leading Koopman eigenfunctions to learn a linear DMD model governing a nonlinear system [285, 90]. To implement EDMD we define a dictionary of $K \geq 1$ user-specified observables

$$\mathcal{D} = \{\psi_1, \psi_2, \ldots, \psi_K\}. \tag{2.47}$$

The goal of EDMD is to approximate the Koopman eigenfunctions in (2.39) as linear combinations of the dictionary elements by obtaining a DMD matrix that advances the observations

Algorithm 2.4 Extended Dynamic Mode Decomposition (EDMD)

1. For each snapshot X_n, gather the subsequent snapshot Y_n
2. Define a dictionary of observable functions $\mathcal{D} = \{\psi_1, \psi_2, \ldots, \psi_K\}$ and construct matrices $\Psi(X)$ as in (2.48) and $\Psi(Y)$ as in (2.49)
3. Compute the DMD matrix $A = \Psi(Y)\Psi(X)^\dagger$
4. Use left eigenvectors $wA = \mu w$ to approximate Koopman eigenfunctions as $w_1\psi_1(x) + w_2\psi_2(x) + \cdots + w_K\psi_K(x)$ with eigenvalue μ

---

forward in time instead of the snapshots. Recalling the snapshot matrices X and Y, we denote the observable data matrices as

$$\Psi(X) = \begin{bmatrix} \psi_1(X_1) & \psi_1(X_2) & \cdots & \psi_1(X_N) \\ \psi_2(X_1) & \psi_2(X_2) & \cdots & \psi_2(X_N) \\ \vdots & \vdots & \ddots & \vdots \\ \psi_K(X_1) & \psi_K(X_2) & \cdots & \psi_K(X_N) \end{bmatrix} \in \mathbb{C}^{K\times N} \tag{2.48}$$

and

$$\Psi(Y) = \begin{bmatrix} \psi_1(Y_1) & \psi_1(Y_2) & \cdots & \psi_1(Y_N) \\ \psi_2(Y_1) & \psi_2(Y_2) & \cdots & \psi_2(Y_N) \\ \vdots & \vdots & \ddots & \vdots \\ \psi_K(Y_1) & \psi_K(Y_2) & \cdots & \psi_K(Y_N) \end{bmatrix} \in \mathbb{C}^{K\times N}. \tag{2.49}$$

EDMD then seeks to solve the minimization problem

$$\underset{A\in\mathbb{C}^{M\times M}}{\arg\min} \; \|\Psi(Y) - A\Psi(X)\|_F^2, \tag{2.50}$$

providing a DMD matrix $A = \Psi(Y)\Psi(X)^\dagger$ that advances the data forward in time *inside* the observables. Much like DMD, we should expect better approximations when K is large, built from a dictionary that is diverse enough to describe the eigenfunctions of the Koopman operator.

With the EDMD matrix A from (2.50) it is now straightforward to compute its spectrum. An eigenvalue $\mu \in \mathbb{C}$ and associated left eigenvector $w = [w_1, w_2, \ldots, w_K] \in \mathbb{C}^K$ generates an approximation of Koopman eigenfunction with the same eigenvalue μ by

$$w_1\psi_1(x) + w_2\psi_2(x) + \cdots + w_K\psi_K(x), \tag{2.51}$$

similar to the analytic case of Example 2.7. Similarly the Koopman modes are approximated using right eigenvectors, arranged as columns of the matrix V. If there exists a vector $b = [b_1, b_2, \ldots, b_K] \in \mathbb{C}^K$ so that the state-space observable is expanded in the dictionary

$$g_m(x) = x_m = b_1\psi_1(x) + b_2\psi_2(x) + \cdots + b_K\psi_K(x) \tag{2.52}$$

for each $m \in \{1, \ldots, M\}$, then the Koopman modes are approximated by $c = (Vb)^\intercal \in \mathbb{C}^K$. This process is summarized in Algorithm 2.4.

EDMD Convergence Properties. EDMD is well studied, and much is known about its ability to approximate the Koopman operator from data. First, if the snapshots X_n are drawn either independently or from an ergodic trajectory, the EDMD matrix A almost surely converges to the L^2 orthogonal projection of $\mathcal{K}$ onto span$\{\psi_1, \ldots, \psi_K\}$ as $N \to \infty$, denoted $\mathcal{K}_\mathcal{D}$ [285]. Furthermore, if the dictionary elements are chosen from a countable orthonormal basis of L^2, then $\mathcal{K}_\mathcal{D} \to \mathcal{K}$ as $K \to \infty$ in the strong operator topology [154]. Similar convergence results hold for the continuous-time and stochastic Koopman operators [42], which are presented in Chapter 4. However, in practice one can only have K and N finite, meaning that the choice of basis functions plays an integral role in the performance of the EDMD algorithm. There is some guidance for the choice of dictionary elements [285] based on experience, while there are also methods of incorporating machine learning to discover near-optimal libraries from data [173, 180, 95].

Example 2.8. We will now apply EDMD to data collected from the mapping (2.40) from Example 2.7. We fix $\lambda_1 = 1.5$ and $\lambda_2 = 0.5$, although other values return nearly identical results. The data matrix $X \in \mathbb{R}^{2\times 10000}$ is built by drawing $N = 10000$ pairs from the uniform distribution on $[-1, 1]^2$, while $Y \in \mathbb{R}^{2\times 10000}$ comes from applying the map (2.40) to each pair in X.

The EDMD dictionary is taken to be all monomials up to degree 2,

$$\mathcal{D}_2 = \{x_1, x_2, x_1^2, x_1x_2, x_2^2\}, \tag{2.53}$$

giving $K = |\mathcal{D}_2| = 5$. Implementing EDMD returns the eigenvalue and approximate Koopman eigenfunction pairs as

$$\begin{aligned}
\underline{\mu = \texttt{0.2254}:} \quad & \texttt{0.0013}x_1 - \texttt{0.5253}x_2 - \texttt{0.0491}x_1^2 + \texttt{0.0056}x_1x_2 - \texttt{0.8495}x_2^2,\\
\underline{\mu = \texttt{0.5000}:} \quad & \texttt{0.0000}x_1 - \texttt{0.9615}x_2 - \texttt{0.2747}x_1^2 + \texttt{0.0000}x_1x_2 + \texttt{0.0000}x_2^2,\\
\underline{\mu = \texttt{0.7503}:} \quad & \texttt{0.5153}x_1 - \texttt{0.0099}x_2 - \texttt{0.0033}x_1^2 + \texttt{0.8570}x_1x_2 + \texttt{0.0064}x_2^2,\\
\underline{\mu = \texttt{1.5000}:} \quad & \texttt{1.0000}x_1 - \texttt{0.0000}x_2 - \texttt{0.0000}x_1^2 + \texttt{0.0000}x_1x_2 + \texttt{0.0000}x_2^2,\\
\underline{\mu = \texttt{2.2500}:} \quad & \texttt{0.0000}x_1 + \texttt{0.0000}x_2 - \texttt{1.0000}x_1^2 + \texttt{0.0000}x_1x_2 + \texttt{0.0000}x_2^2.
\end{aligned} \tag{2.54}$$

Notice that the three eigenpairs from (2.44) are represented in (2.54). However, since $K = 5$, there are five eigenvalues, leading to two "spurious" eigenvalues: 0.2254 and 0.7503. Prior to explaining what these values represent, let us experiment with a larger dictionary first.

If we increase our dictionary to include all monomials in x_1 and x_2 up to degree 3, i.e.,

$$\mathcal{D}_3 = \{x_1, x_2, x_1^2, x_1x_2, x_2^2, x_1^3, x_1^2x_2, x_1x_2^2, x_2^3\}, \tag{2.55}$$

now $K = |\mathcal{D}_3| = 9$. Thus, we expect 9 eigenvalues of the EDMD matrix, while Example 2.7 has shown that the Koopman operator only requires three of them to linearize (2.40). Implementing EDMD with the same snapshot matrices X and Y from before, of the resulting 9 eigenvalues, exactly 3 of them are the expected $\lambda_1, \lambda_2, \lambda_1^2$ that lead to the Koopman eigenfunctions from Example 2.7. Of the remaining 6, one eigenvalue is $\mu = \texttt{0.7500}$ with eigenfunction

$$\texttt{0.96152}x_1x_2 + \texttt{0.27472}x_1^3. \tag{2.56}$$

We have neglected all terms with coefficients smaller in absolute value than 10^{-4} for clarity. The function (2.56) is approximately $\psi_{\lambda_1}\psi_{\lambda_2}$ with eigenvalue approximately $\lambda_1\lambda_2$. This is no coincidence since one can check that for the system (2.40), using the general Koopman operator formulation (2.37) we indeed have

$$\mathcal{K}\psi_{\lambda_1}\psi_{\lambda_2} = [\psi_{\lambda_1}\psi_{\lambda_2}] \circ F = [\psi_{\lambda_1} \circ F] \cdot [\psi_{\lambda_2} \circ F] = \lambda_1\lambda_2\psi_{\lambda_1}\psi_{\lambda_2}, \tag{2.57}$$

giving that $\lambda_1\lambda_2$ is another eigenvalue of the Koopman operator with eigenfunction $\psi_{\lambda_1}\psi_{\lambda_2}$. Hence, the presence of the eigenvalue $\mu = \texttt{0.7503}$ using the dictionary $\mathcal{D}_2$ is not spurious at all, as it approximates an eigenvalue of the Koopman operator. However, the dictionary $\mathcal{D}_2$ was insufficient to describe its eigenfunction, leading its projection onto this dictionary instead.

One can follow the arguments in (2.57) to show that if ψ and ϕ are a Koopman eigenfunction with eigenvalue μ and ν, then for any $p, q \in \mathbb{R}$ for which $\psi^p\phi^q$ belongs to the specified space of observable functions, $\psi^p\phi^q$ is also an eigenfunction with eigenvalue $\mu^p\nu^q$. This statement holds in general for the Koopman operator and not just for the current example.

Turning back to the system (2.40), the EDMD algorithm also outputs the eigenvalue $\mu = \texttt{3.3750} \approx \lambda_1^3$ for which the eigenfunction approximates the true Koopman eigenfunction x_1^3 with eigenvalue λ_1^3. Unfortunately, this does not explain the presence of the eigenvalue $\mu = \texttt{3.3750}$ over any of the other infinitely many choices for Koopman eigenvalues to be represented. However, one can readily check that the other eigenvalues that arise from EDMD can be written as $\lambda_1^p\lambda_2^q$, although the exact values of p and q that are represented in numerical experiments highly depend on the data and dictionary used to seed EDMD.

2.5 ▪ The Kernel Trick

We saw that EDMD seeks to approximate Koopman eigenfunctions using only data gathered from a system. At the core of the method is the crucial step of defining a dictionary to expand the eigenfunctions in. Larger dictionaries provide a more robust basis to more accurately describe the eigenfunctions. A drawback arises in the case that the dimension of the space from which our data is drawn, $\mathbb{R}^M$, is large. To see why, suppose we construct a dictionary of polynomials up to degree $d \geq 1$ in our $M \geq 1$ variables. The number of polynomial terms at degree j is given by $\binom{M+j-1}{j}$, and summing from $j = 1$ to $j = d$ gives

$$K := |\mathcal{D}| = \sum_{j=1}^{d} \binom{M+j-1}{j} = \binom{M+d}{d} - 1, \tag{2.58}$$

where the 1 being subtracted in the final answer comes from the fact that we did not include constant functions in the dictionary. Recalling Example 2.1, where the spatial domain of the NLS is divided into $M = 64$ points, (2.58) would give $K = 2145$ using $d = 2$, $K = 47905$ using $d = 3$, and $K = 814385$ using $d = 4$. In practice we should expect to have $d \gg 1$ to get the best approximations, but the computation quickly becomes intractable, as we are expected to construct, multiply, and find pseudoinverses and eigenvalues of extremely large matrices. It is exactly this issue that **kernel DMD** methods seek to overcome. That is, while EDMD works well when the dimension of the state space is small, kernel methods provide an alternative for approximating Koopman eigenfunctions when the dimension of the state space is large.

We follow the presentation of kernel DMD that was proposed by Williams, Rowley, and Kevrekidis [286]. We begin with a dictionary $\mathcal{D}$ and define the resulting EDMD matrix $A = \Psi(Y)\Psi(X)^\dagger$, formed from the data matrices (2.48) and (2.49). Recall that we are interested in the spectrum of A since it approximates the spectrum of the Koopman operator. Supposing that we have the (reduced) SVD of $\Psi(X)$, we get $\Psi(X) = U\Sigma V^\intercal$, where $U \in \mathbb{R}^{K\times N}$ while $\Sigma, V \in \mathbb{R}^{N\times N}$. Since the range of $A^\intercal$ is a subset of the range of $\Psi(X)$, a left eigenvector w of A with eigenvalue $\mu \neq 0$ can be written $w = \hat{w}U^\intercal$. Then, from $\mu w = wA$ we get

$$\begin{aligned} \mu\hat{w}U^\intercal &= \mu w \\ &= wA \\ &= \hat{w}U^\intercal\Psi(Y)\Psi(X)^\dagger \\ &= \hat{w}U^\intercal\Psi(Y)V\Sigma^\dagger U^\intercal \\ &= \hat{w}\Sigma^\dagger V^\intercal\Psi(X)^\intercal\Psi(Y)V\Sigma^\dagger U^\intercal. \end{aligned} \tag{2.59}$$

The above means that $\hat{w}$ is an eigenvalue of $\hat{A} := \Sigma^\dagger V^\intercal \Psi(X)^\intercal \Psi(Y) V \Sigma^\dagger \in \mathbb{R}^{N\times N}$ with eigenvalue μ as well. Notice that the dimension of $\hat{A}$ is now based on the number of snapshots instead of the number of dictionary elements. Thus, in the case when $K \gg N$ we have effectively reduced the dimension of the problem since a left eigenvector $\hat{w}$ of $\hat{A}$ leads to a left eigenvector $w = \hat{w}U^\intercal$ of A with the same eigenvalue, while the major bottleneck of finding eigenvalues and eigenvectors is reduced from a $K \times K$ matrix to one that is $N \times N$.

Now, the SVD of $\Psi(X)$ further gives that

$$\Psi(X)^\intercal \Psi(X) = V\Sigma^2 V^\intercal, \tag{2.60}$$

which, since V is an orthogonal matrix, represents the diagonalization of the matrix $\Psi(X)^\intercal\Psi(X)$. To compute $\hat{A}$ from data we can equivalently find the eigendecomposition of $\Psi(X)^\intercal\Psi(X)$ to obtain Σ and V. In principle, this should be a simpler task than obtaining the SVD of $\Psi(X)$ since $\Psi(X)^\intercal\Psi(X)$ is an $N \times N$ matrix, but constructing $\hat{A}$ still requires defining the library elements and evaluating them at each snapshot X_n. When $K \gg 1$ we require a significant number of computations just to produce the matrix $\Psi(X)^\intercal\Psi(X)$. The goal of kernel DMD is to build $\Psi(X)^\intercal\Psi(X)$ without explicitly defining the observables that make up the dictionary.

To circumvent defining the dictionary $\mathcal{D}$, we implement what is known as the **kernel trick**. First, notice that entries of the matrix $\Psi(X)^\intercal\Psi(X)$ are given by

$$\begin{aligned}[\Psi(X)^\intercal\Psi(X)]_{i,j} &= \begin{bmatrix}\psi_1(X_i) & \cdots & \psi_K(X_i)\end{bmatrix} \cdot \begin{bmatrix}\psi_1(X_j) & \cdots & \psi_K(X_j)\end{bmatrix} \\ &=: \psi(X_i)^\intercal\psi(X_j),\end{aligned} \tag{2.61}$$

thus representing an inner product between the vectors of observable functions evaluated at snapshots X_i and X_j with $1 \le i,j \le N$. The kernel trick relies on a **kernel function** $h : \mathbb{R}^M \times \mathbb{R}^M \to \mathbb{R}$ so that $h(X_i, X_j) = \psi(X_i)^\intercal\psi(X_j)$ for all $1 \le i,j \le n$ [51]. By defining a specific kernel h one is effectively defining the ψ_k in the EDMD dictionary $\mathcal{D}$ without computing these inner products directly. Let us illustrate with the following example.

Example 2.9. For exposition take $M = 2$, and define the polynomial kernel $h(x,y) = (1 + y^\intercal x)^d$ for each $x = (x_1, x_2), y = (y_1, y_2) \in \mathbb{R}^2$ of degree $d \ge 1$. When $d = 2$ we have

$$\begin{aligned} h(x,y) &= (1 + y^\intercal x)^2 \\ &= (1 + x_1y_1 + x_2y_2)^2 \\ &= 1 + 2x_1y_1 + 2x_2y_2 + 2x_1x_2y_1y_2 + x_1^2y_1^2 + x_2^2y_2^2. \end{aligned} \tag{2.62}$$

The above can be recognized as an inner product of the form (2.61) involving

$$\psi(x) = \begin{bmatrix}1 & \sqrt{2}x_1 & \sqrt{2}x_2 & x_1^2 & x_1x_2 & x_2^2\end{bmatrix}, \tag{2.63}$$

thus representing a dictionary of monomials in two variables up to degree 2. One can similarly check that taking $d = 3$ and expanding the kernel results in an inner product of the form (2.61) using a dictionary of monomials up to degree 3. In general, the polynomial kernel $h(x,y) = (1 + y^\intercal x)^d$ for $x, y \in \mathbb{R}^M$ for any $M \ge 1$, and $d \ge 1$ produces an equivalent inner product to using a dictionary of monomials in M variables up to degree d.

Coupling the above example of the polynomial kernel with the preceding discussion provides a computationally tractable way to perform EDMD with high-dimensional data. Indeed, we can build the matrix $\Psi(X)^\intercal\Psi(X)$ using a dictionary of monomials in M variables up to degree d with

$$[\Psi(X)^\intercal\Psi(X)]_{i,j} = (1 + X_j^\intercal X_i)^d, \tag{2.64}$$

Algorithm 2.5 Kernel DMD

1. For each snapshot X_n, gather the subsequent snapshot Y_n
2. Using a kernel function $h(x, y)$ compute matrices $\Psi(X)^\intercal\Psi(X) = [h(X_i, X_j]_{i,j}$ and $\Psi(X)^\intercal\Psi(Y) = [h(X_i, Y_j)]_{i,j}$
3. Diagonalize $\Psi(X)^\intercal\Psi(X) = V\Sigma^2V^\intercal$
4. Use V, Σ, and $\Psi(X)^\intercal\Psi(Y)$ to compute $\hat{A} = \Sigma^\dagger V^\intercal\Psi(X)^\intercal\Psi(Y)V\Sigma^\dagger$
5. Use left eigenvectors $\hat{w}\hat{A} = \mu\hat{w}$ to approximate Koopman eigenfunctions as $\hat{w}\Sigma^\dagger V^\intercal \begin{bmatrix} h(X_1, x) & h(X_2, x) & \cdots & h(X_N, x)\end{bmatrix}^\intercal$ with eigenvalue μ

for which the complexity of each computation does not increase with the degree d. We can also build the matrix $\Psi(X)^\intercal\Psi(Y)$ using the kernel, meaning that we may produce $\hat{A}$ using kernels.

Other commonly encountered kernels are the Gaussian kernel

$$h(x, y) = \mathrm{e}^{-\|x-y\|^2/\sigma^2} \tag{2.65}$$

and the Laplace kernel

$$h(x, y) = \mathrm{e}^{-\|x-y\|/\sigma}, \tag{2.66}$$

each involving a tunable parameter $\sigma > 0$. As with the choice of the dictionary elements for EDMD, it is not clear what the optimal choice of kernel is for a given problem. Furthermore, general kernels do not lead to finite-dimensional dictionaries in the same way that the polynomial kernel does, and therefore it is often not possible to compare the results of kernel DMD to EDMD.

Now that we have identified a method to compute $\hat{A}$ using kernel functions, it remains to use this information to approximate the Koopman eigenfunctions. Recall from (2.51) that eigenfunctions of the Koopman operator are approximated by $w\psi(x)^\intercal$, where w is a left eigenvector of the EDMD matrix $\Psi(Y)\Psi(X)^\dagger$. If $\hat{w}$ is a left eigenvector of $\hat{A}$, from above we know that $w = \hat{w}U^\intercal$ is a left eigenvector of $\Psi(Y)\Psi(X)^\dagger$. The eigenfunctions are then given by

$$\begin{aligned} w\psi(x)^\intercal &= \hat{w}U^\intercal\psi(x)^\intercal \\ &= \hat{w}\Sigma^\dagger V^\intercal\Psi(X)^\intercal\psi(x)^\intercal \\ &= \hat{w}\Sigma^\dagger V^\intercal \begin{bmatrix} h(X_1, x) & h(X_2, x) & \cdots & h(X_N, x)\end{bmatrix}^\intercal, \end{aligned} \tag{2.67}$$

thus giving a representation entirely in terms of the computed matrices Σ, V and the kernel function h. If we denote $\mathcal{I}_m(X) \in \mathbb{R}^N$ as the vector of all snapshots of the state-space observable x_m, then its corresponding Koopman modes are recovered by $\hat{Q}^\intercal\Sigma^\dagger V^\intercal\mathcal{I}_m(X)$, where $\hat{Q}$ is a matrix of right eigenvectors of $\hat{A}$ [286]. The process of using kernels to approximate the Koopman operator is summarized in Algorithm 2.5.

Example 2.10. To implement kernel DMD, we consider the scalar Burgers equation

$$u_t = \nu u_{xx} - uu_x, \tag{2.68}$$

where $\nu > 0$ is the viscosity parameter. We restrict $x \in [-L/2, L/2]$ for some $L > 0$ and impose Dirichlet boundary conditions

$$u(\pm L/2, t) = 0 \qquad \forall t \geq 0. \tag{2.69}$$

While the Burgers equation (2.68) is a nonlinear PDE, there exists a change of variable that recasts it as the linear heat equation, which is completely solvable on simple domains. Thus, the Burgers equation is one of few examples of a nonlinear PDE that is analytically solvable. This solvability can therefore be leveraged to understand its associated Koopman operator.

For a solution $u(x,t)$ to (2.68), we introduce the new scalar function $v(x,t)$ given by

$$v(x,t) := \frac{\mathrm{e}^{\xi(u(x,t))}}{\int_{-L/2}^{L/2} \mathrm{e}^{\xi(u(x,t))}\mathrm{d}x}, \tag{2.70}$$

where $\xi(u) = -\frac{1}{2\nu}\int_{-L/2}^{x} u(y,t)\mathrm{d}y$. The transformation of u into v is referred to as the Cole–Hopf transformation, and it provides that v satisfies

$$v_t = \nu v_{xx}, \tag{2.71}$$

the linear heat equation. Furthermore, the Cole–Hopf transformation is invertible, so a solution v of (2.71) gives a solution to the Burgers equation via

$$u(x,t) = -2\nu v_x/v. \tag{2.72}$$

Finally, the Dirichlet boundary conditions on u endows v with Neumann boundary conditions, i.e., $v_x(\pm L/2, t) = 0$ for all $t \geq 0$, again through the Cole–Hopf transformation (2.70).

As was demonstrated in [211], the Cole–Hopf transformation can be used to understand the Koopman operator associated to the Burgers equation. To begin, the prescribed Neumann boundary conditions on (2.71) give a sequence of eigenfunctions

$$w_\ell(x) = \begin{cases} \cos\left(\frac{\ell\pi x}{L}\right), & \ell \text{ even}, \\ \sin\left(\frac{\ell\pi x}{L}\right), & \ell \text{ odd}, \end{cases} \tag{2.73}$$

for each $\ell \geq 0$. We can therefore write the solution $v(x,t)$ to (2.71) as

$$v(x,t) = \sum_{\ell=0}^{\infty} \phi_\ell(v(x,t)) w_\ell(x), \tag{2.74}$$

where $\phi_\ell(v(x,t)) := \frac{2}{L}\int_{-L/2}^{L/2} w_\ell(x)v(x,t)\mathrm{d}x$ is the standard projection onto each eigenfunction w_ℓ. Importantly, one can interpret the ϕ_ℓ as observables of the solution to the heat equation v. Using separation of variables we know that each observable evolves in time according to

$$\phi_\ell(v(x,t)) = \phi_\ell(v(x,0))\mathrm{e}^{-\frac{\ell^2\pi^2\nu t}{L^2}}, \tag{2.75}$$

giving that ϕ_ℓ are eigenfunctions of the (continuous-time) Koopman operator of the linear heat equation with Koopman eigenvalue $-\ell^2\pi^2\nu/L^2$.

One should be careful in interpreting the above since observables in the PDE case are functions whose input is a solution to the heat equation (2.71), meaning their domain is an infinite-dimensional space of functions over (x,t). This is a nontrivial extension of the Koopman operator since we know the Koopman modes in the eigenfunction expansion (2.38) are functions of the spatial domain x. In the coming computational application we discretize space into M elements, and so defining the vector $[v_m]_{m=1}^{M} \in \mathbb{R}^M$ by $v_m := v(x_m, t)$ for some $t \geq 0$ and evaluating (2.74) at $x = x_m$ provides a Koopman eigenfunction expansion for the state-space observable v_m with corresponding modes $w_\ell(x_m)$.

In [211] the Cole–Hopf transformation is employed to understand the Koopman eigenfunction expansion of solutions to the Burgers equation. We highlight these findings here. One can use the inverse transformation (2.72) to define a recurrence relation for the Koopman eigenfunctions, eigenvalues, and modes in the Koopman expansion of $u(x,t)$. The result is that we can write

$$u(x,t) = \sum_{\ell=1}^{\infty} \Phi_\ell(u)\hat{w}_\ell(x)\mathrm{e}^{-\frac{\ell^2\pi^2\nu t}{L^2}}, \tag{2.76}$$

where $\Phi_\ell(u)$ is the eigenfunction, $\hat{w}_\ell(x)$ is the mode, and the (continuous-time) eigenvalues are $-\ell^2\pi^2\nu/L^2$. The first three terms in the sum are given by

$$\begin{aligned}
\Phi_1(u)\hat{w}_1 &= -\frac{2\pi\nu}{L}\left(\frac{\phi_1(v(u))}{\phi_0(v(u))}\right)\cos\left(\frac{\pi x}{L}\right), \\
\Phi_2(u)\hat{w}_2 &= \frac{\pi\nu}{L}\left(\frac{\phi_1(v(u))}{\phi_0(v(u))}\right)^2\sin\left(\frac{2\pi x}{L}\right), \\
\Phi_3(u)\hat{w}_3 &= -\frac{2\pi\nu}{L}\left(\frac{\phi_1(v(u))}{\phi_0(v(u))}\right)^3\sin^2\left(\frac{\pi x}{L}\right)\cos\left(\frac{\pi x}{L}\right),
\end{aligned} \tag{2.77}$$

where $v(u)$ is meant to cast v as a function of u using the Cole–Hopf transformation (2.70). Precisely, $v(u)$ is an observable of the Burgers equation which recasts the dynamics as linear.

With the theory now covered, we implement kernel DMD on simulated snapshot data from the Burgers equation. Although our analytical work derived a Koopman expansion of the solution to the Burgers equation in continuous time, with a sufficient sampling rate $h > 0$ we should expect

$$u(x,nh) \approx \sum_{\ell=1}^{\infty} \Phi_\ell(u)\hat{w}_\ell(x)\left(\mathrm{e}^{-\frac{\ell\pi^2\nu h}{L^2}}\right)^n, \quad n \geq 0. \tag{2.78}$$

This constitutes an approximate Koopman expansion for the discrete-time system that advances between solutions of the Burgers equation at $t = nh$ and $t = (n+1)h$. Thus, we expect the Koopman eigenvalues of this discrete-time process to be approximately $\mathrm{e}^{-\ell^2\pi^2\nu h/L^2}$ when h is small. With this in mind, we generate snapshots of Burgers equation (2.68) with $(\nu, L) = (0.1, 2\pi)$ over a spatial medium into $M = 256$ evenly spaced points. Snapshots are obtained from numerical integration of five initial conditions taking the form

$$u(x,0) = \sum_{j=1}^{J} a_j \sin(jx), \tag{2.79}$$

where $1 \leq J \leq 10$ is a random integer and the a_j are drawn independently from the uniform distribution on $[-1,1]$. Initial conditions are integrated from $t = 0$ to $t = 10$, while snapshot spacing is taken to be $h = 0.1$. In total we have $N = 500$. An example solution integrated in t is provided in Figure 2.10(a), where one sees its decay to zero, as guaranteed by the Koopman expansion (2.76).

We report results using the polynomial kernel of degrees $d = 2, 4, 6, 8$. Eigenvalues of the reduced DMD matrix $\hat{A} \in \mathbb{R}^{500\times 500}$ are plotted in the complex plane in Figure 2.10(b)–(e). From (2.78) we expect that the discrete-time Koopman operator has only real, positive eigenvalues. This is not the case for our numerical results; however, as d increases we do have eigenvalues coalescing onto the unit interval along the real axis. Recall from the previous section that convergence could be improved by increasing the number of snapshots N, but this may come at the cost of a computational bottleneck. Furthermore, we saw in Example 2.8 that there

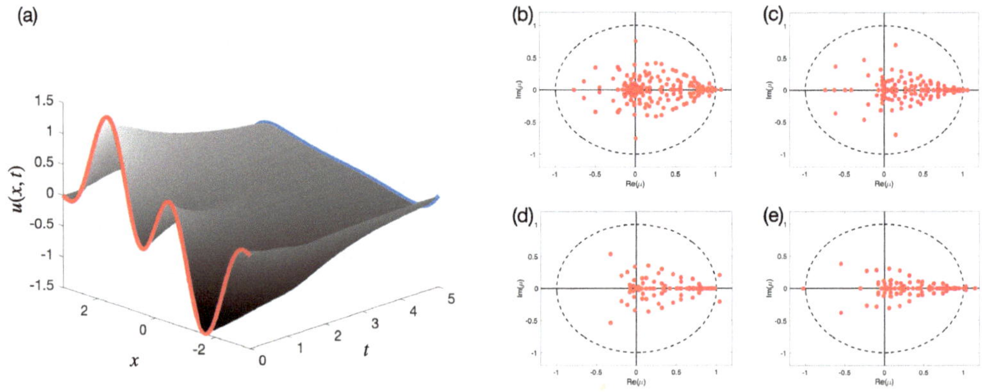

Figure 2.10. *The results of kernel DMD applied to snapshots of the Burgers equation.* (a) *A random initial condition (red) of the form* (2.79) *decays in time to be nearly identically zero (blue) by* $t = 5$. (b)–(e) *The eigenvalues of kernel DMD using the polynomial kernel* $h(x, y) = (1 + y^\intercal x)^d$ *for* $d = 2, 4, 6, 8$, *respectively.*

are uncountably many Koopman eigenvalues that the eigenvalues of $\hat{A}$ could be approximating, and there is almost no way of knowing which will be selected. Therefore, we should not expect to see convergence as $N \to \infty$ to the eigenvalues $\mathrm{e}^{-\ell^2\pi^2\nu h/L^2}$, $\ell \geq 1$, in the eigenvalues of $\hat{A}$.

Finally, the Cole–Hopf transformation is an illustration of all that EDMD and the Koopman operator hope to achieve for nonlinear systems. That is, the Cole–Hopf transformation makes the Burgers equation linear and can be solved using separation of variables. If we performed EDMD with $v(u)$ being an observable in the dictionary, one would expect to get a subset of the approximate discrete-time eigenvalues $\mathrm{e}^{-\ell^2\pi^2\nu h/L^2}$. In the accompanying script for this section, we demonstrate this by taking only one solution to the Burgers equation with initial condition (2.79), applying the Cole–Hopf transformation, and learning a DMD model on the transformed variables. In this case the EDMD matrix is size 256×256, and a subset of its eigenvalues can be seen to closely correspond to the discrete-time Koopman eigenvalues with small values of ℓ. Results of this application of EDMD can be improved by including longer snapshots and those integrated from different initial conditions. Similar results are reported in [211, 21]. The drawback is that one requires the Cole–Hopf transformation to achieve such impressive results, something which can only be known from the governing equations of the system.

2.6 ▪ Extensions and Applications of DMD

In this chapter we were introduced to powerful methods to obtain linear descriptions of nonlinear processes. All of these methods are broadly organized around DMD, which itself is a method inspired by the Koopman operator. As we saw in this chapter, techniques such as EDMD offer a way of approximating the Koopman operator using only matrix algebra. We should constantly be reminded that although there is a dense mathematical theory that underlies much of what we have done in this section, implementing these ideas requires little more computation than matrix multiplication and (pseudo) inversion. These operations are some of the earliest methods to be implemented on a computer, and therefore there are quick and efficient algorithms that implement them, even on large matrices. Furthermore, the use of the SVD in finding DMD matrices naturally allows for a dimensionality reduction by projecting onto only the largest singular values [111], which can be used to ease some of the computation in extremely high-dimensional scenarios. For a review of dimensionality reduction and DMD, see [116].

While the organization of this chapter was primarily built around applying the methods to dynamical systems, DMD is useful for understanding and interpreting real-world data as well. For example, DMD has been used to separate background and foreground components in video data [107, 114] and also to develop financial trading strategies [189], among other successful applications that are detailed in the review [10]. With real-world data comes a number of issues that were not addressed in this chapter, including noisy measurements, irregularly spaced snapshots, and disentangling external inputs to the system. Filtering methods have been proposed to obtain DMD matrices with regularly spaced noisy measurements [208], and [229] provides a method to disambiguate control inputs from the system dynamics. To overcome irregularly spaced data in time Askham and Kutz introduced a nonconvex algorithm [19] that performs well in the presence of additive noise, which was extended in [172] to handle multiplicative noise too.

2.7 ▪ Appendix: The Continuous-Time Koopman Operator

In Example 2.10 we were introduced to the continuous-time Koopman operator. Here we make this introduction precise. We begin with the dynamical system

$$\dot{x} = F(x), \tag{2.80}$$

where $x = x(t) \in \mathcal{M}$ and $x(0) = x_0 \in \mathcal{M}$. Under mild conditions on F we may define the flow operator $\Phi : \mathbb{R}_+ \times \mathcal{M} \to \mathcal{M}$ so that a solution of (2.80), $x(t)$, with initial condition $x(0) = x_0$ is given by $x(t) = \Phi(t, x_0)$. Using the flow operator we define the time-h map $F_h(x) := \Phi(h, x)$, which pushes any solution to (2.80) forward by $h > 0$ time units. Since $F_h : \mathcal{M} \to \mathcal{M}$ now generates a discrete dynamical system, we consider its associated Koopman operator, denoted $\mathcal{K}_h$. The continuous-time Koopman operator for (2.80) is a semigroup that acts on continuously differentiable observables $\phi : \mathcal{M} \to \mathbb{C}$ with infinitesimal generator $\mathcal{L}$ given by

$$\mathcal{L}\phi := \lim_{h \to 0^+} \frac{\phi \circ F_h - \phi}{h} = F \cdot \nabla\phi. \tag{2.81}$$

Formally, the resulting Koopman operator is the linear operator $\mathcal{K}_t\phi := \mathrm{e}^{\mathcal{L}t}\phi$ for each $t \geq 0$. Here $\mathcal{L}$ is called the **Lie derivative** of the flow (2.80) and will be a focal point of Chapter 4.

Eigenpairs (μ, ϕ_μ) of the Lie derivative $\mathcal{L}$ are solutions of the PDE

$$F \cdot \nabla\phi_\mu = \mu\phi_\mu, \tag{2.82}$$

which are invariant under the Koopman operator $\mathcal{K}_t$ in that

$$\mathcal{K}_t\phi_\mu(x) = \mathrm{e}^{\mu t}\phi_\mu(x) \tag{2.83}$$

for each $x \in \mathcal{M}$ and all $t \geq 0$. Notice that evaluating at $t = h$ gives

$$\mathcal{K}_h\phi_\mu(x) = \mathrm{e}^{\mu h}\phi_\mu(x), \tag{2.84}$$

meaning that $(\mathrm{e}^{\mu h}, \phi_\mu)$ is an eigenpair of the Koopman operator $\mathcal{K}_h$.

Collecting data from (2.80) with small sampling rate $h > 0$ allows one to apply EDMD and obtain an approximation of $\mathcal{K}_h$. Following as in Section 2.4, eigenpairs of the EDMD matrix, denoted K_h, approximate eigenpairs of (2.84), which can in turn approximate the continuous-time eigenpairs via (2.83). Furthermore, we use K_h to approximate the Lie derivative $\mathcal{L}$, denoted L, acting on the dictionary using finite differences

$$L \approx \frac{K_h - I}{h} \tag{2.85}$$

Algorithm 2.6 Approximating Koopman Eigenfunctions in Continuous Time

1. Gather snapshot pairs (X_n, Y_n) with Y_n the state of X_n exactly $h > 0$ time units in the future

2. Define a dictionary of observable functions $\mathcal{D} = \{\psi_1, \psi_2, \ldots, \psi_K\}$ and construct matrices $\Psi(X)$ as in (2.48) and $\Psi(Y)$ as in (2.49)

3. Compute the Koopman matrix $K_h = \Psi(Y)\Psi(X)^\dagger$

4. Approximate the Lie derivative $L_h = [K_h - I]/h$

5. Use left eigenvectors $wL_h = \mu w$ to approximate continuous-time Koopman eigenfunctions as $w_1\psi_1(x) + w_2\psi_2(x) + \cdots + w_K\psi_K(x)$ with eigenvalue μ

or, following [191], the matrix logarithm

$$L \approx \frac{1}{h} \log(K_h). \tag{2.86}$$

Thus, although the continuous and discrete formulations of the Koopman operator differ, the approach to approximating them numerically is largely the same. This can be seen by comparing the similarity of the approach in this section summarized in Algorithm 2.6 and discrete-time EDMD summarized in Algorithm 2.4.

Chapter 3

Identifying Nonlinear Dynamics

In the previous chapter we saw that DMD and its variants produce linear models to forecast and interpret nonlinear dynamics. Although DMD is a powerful method in data analysis and dynamical systems, the models it produces often lack the ability to generalize beyond the training set. To see this for yourself, use the DMD matrix from Example 2.1 to forecast a different initial state of the NLS (2.8), and compare it to the result from integrating the system with the same initial condition. This will almost certainly result in a DMD prediction that quickly diverges from the true dynamics of the system. This should be expected since the NLS is a nonlinear equation, while DMD produces a linear system based only on the training data given to it. To produce models that can forecast and interpret outside of the training set, we require nonlinearity.

In this chapter we will focus on fitting data to nonlinear models that can be generalized across a range of initial conditions. We will employ sparse regression to fit the models, with the goal of demonstrating the effect of different dictionaries and optimization parameters. We further explore how system identification methods can be used to not only discover new models from data but also complement and improve our understanding those that we already have.

3.1 ▪ Sparse Identification of Nonlinear Dynamics

We begin with a method to produce accurate nonlinear models from data, termed the *Sparse Identification of Nonlinear Dynamics* (SINDy) method [50]. SINDy frames model discovery as a sparse regression problem and seeks to identify an accurate and parsimonious dynamical system to fit time-series data. Like EDMD, it fits models to a dictionary (or library) of candidate functions that are user-prescribed. In fact, EDMD and SINDy are preforming a similar task, with the former being interested in how the system evolves inside of arbitrary observables, while the latter is only concerned with the evolution of the "state-space observables" (2.38). Thus, EDMD and SINDy should be seen as complementary methods of model discovery that focus on different aspects of the same problem, and extensions of one can often be applied to the other as well.

The setting for SINDy is similar to DMD. We begin with data matrices $X, Y \in \mathbb{R}^{M\times N}$, whose columns represent snapshots of the data and rows the number of components to the data. The columns of Y can be as in DMD, representing the next iterate of the respective columns in X if the underlying dynamics are discrete, or, alternatively, they can be estimates of the time derivative of the columns in X at the same time if the underlying dynamics are continuous. We consider a dictionary

$$\mathcal{D} = \{\theta_1, \theta_2, \ldots, \theta_K\}, \qquad K \geq 1, \tag{3.1}$$

with $\theta_k : \mathbb{R}^M \to \mathbb{R}$ for each $1 \le k \le K$. Writing $X = [X_1\ X_2\ \cdots\ X_N]$ with the X_n being the snapshots of X, we denote the matrix

$$\Theta(X) = \begin{bmatrix} \theta_1(X_1) & \theta_1(X_2) & \cdots & \theta_1(X_N) \\ \theta_2(X_1) & \theta_2(X_2) & \cdots & \theta_2(X_N) \\ \vdots & \vdots & \ddots & \vdots \\ \theta_K(X_1) & \theta_K(X_2) & \cdots & \theta_K(X_N) \end{bmatrix} \in \mathbb{R}^{K\times N} \tag{3.2}$$

so that the columns of $\Theta(X)$ represent the snapshots evaluated at each of the K elements of the dictionary. Similar to DMD, model discovery seeks to solve the minimization problem

$$\operatorname*{arg\,min}_{\Xi\in\mathbb{R}^{M\times K}} \|Y - \Xi\Theta(X)\|_F^2. \tag{3.3}$$

The minimizer $\Xi = Y\Theta(X)^\dagger \in \mathbb{R}^{M\times K}$ of (3.3) is a matrix of the coefficients of the linear combination of dictionary elements that best describe the nonlinear correspondence from the columns in $\Theta(X)$ to the columns in Y. In discrete-time systems we have the nonlinear iteration scheme

$$x_{n+1} = \Xi_1\theta_1(x_n) + \Xi_2\theta_2(x_n) + \cdots + \Xi_K\theta_K(x_n), \tag{3.4}$$

while in continuous-time systems we have the ODE

$$\dot{x} = \Xi_1\theta_1(x) + \Xi_2\theta_2(x) + \cdots + \Xi_K\theta_K(x), \tag{3.5}$$

where Ξ_k denotes the kth column of Ξ. Model identification in this way is essentially following EDMD to determine the coefficients in the spans (2.38) to generate the state-space observables.

Although the solution to (3.3) provides the best fit of a nonlinear model between the data in X and Y, it may not be the best from a practical standpoint. First, we want to have a large dictionary that can describe diverse dynamical phenomena. However, the solution Ξ to (3.3) is in general not sparse, meaning that the discovered nonlinear function $\Xi\Theta : \mathbb{R}^M \to \mathbb{R}^M$ will be quite complicated since it uses nearly every element of a large dictionary. Such a model is hardly interpretable and may defeat the purpose of model discovery in the first place. Furthermore, small corruptions of the data through noisy measurements or estimations of the numerical derivative to build Y can lead to small, nonzero coefficients and introduce spurious dictionary elements into the discovered model. The SINDy method seeks to counter these issues by solving (3.3) with an additional sparsity constraint on the matrix Ξ. That is, we seek a description of the nonlinear dynamics that uses as few dictionary terms as possible.

The minimization problem (3.3) is therefore augmented to promote sparsity in Ξ through

$$\operatorname*{arg\,min}_{\Xi\in\mathbb{R}^{M\times K}} \|Y - \Xi\Theta(X)\|_F^2 + \nu\|\Xi\|_0. \tag{3.6}$$

Here $\|\cdot\|_0$ is the "0-norm," which returns a count of the nonzero entries in a matrix, while $\nu > 0$ is a tuning parameter. Larger ν preferences sparsity in Ξ, while smaller ν puts preference on fitting the data with $\|Y - \Xi\Theta(X)\|_F^2$. Finding solutions to sparse regression problems like (3.6) is known to be NP-hard [205, 69], with techniques such as LASSO replacing the 0-norm with the 1-norm, $\|\cdot\|_1$, to make the problem tractable [265, 272]. However, minimizing with the 1-norm gives "approximately" sparse solutions; i.e., there are many near zero terms but not exactly zero.

The SINDy method presents an alternative relaxation of the sparse regression problem (3.6). Define a **sparsity parameter** $\lambda > 0$. Then, letting $Y\Theta(X)^\dagger = \Xi = [\xi_{m,k}]$ be the least squares solution of (3.3), we proceed to minimize

$$\operatorname*{arg\,min}_{\tilde{\Xi}\in\mathbb{R}^{M\times K}} \|Y - \tilde{\Xi}\Theta(X)\|_F^2, \tag{3.7}$$

Algorithm 3.1 Sparse Identification of Nonlinear Dynamics (SINDy)

1. Gather system snapshots X_n and derivative approximations Y_n at each snapshot
2. Define a dictionary $\mathcal{D} = \{\theta_1, \theta_2, \dots, \theta_K\}$ and construct the matrix $\Theta(X)$ as in (3.2)
3. Define a threshold $\lambda > 0$ and perform sequential least-squares:
   (a) Obtain the matrix $\Xi = [\xi_{m,k}]_{m,k} = Y\Theta(X)^\dagger$
   (b) Obtain minimizer $\tilde{\Xi} = [\tilde{\xi}_{m,k}]_{m,k}$ of $\|Y - \tilde{\Xi}\theta(X)\|_F^2$ subject to $\tilde{\xi}_{m,k} = 0$ if $|\xi_{m,k}| < \lambda$
   (c) Repeat (b) by replacing $\tilde{\Xi}$ with Ξ until these matrices are the same
4. Converged coefficient matrix $\Xi = [\Xi_1 \dots \Xi_K]$ approximates the ODE $\dot{x} = \Xi_1\theta_1(x) + \Xi_2\theta_2(x) + \cdots + \Xi_K\theta_K(x)$

with the restriction that the new coefficient matrix $\tilde{\Xi} = [\tilde{\xi}_{m,k}] \in \mathbb{R}^{M \times K}$ has $\tilde{\xi}_{m,k} = 0$ if $|\xi_{m,k}| \le \lambda$. That is, we assign zero coefficients to all elements of the dictionary for which a coefficient in Ξ is smaller than λ in absolute value. By manually enforcing a sparsity pattern in (3.7) we arrive at another Procrustes problem.

Upon solving (3.7), we can repeat this manual sparsity enforcement to continue to eliminate terms from our nonlinear model. Precisely, letting $\tilde{\Xi}$ take the place of Ξ above, we then use the sparsity parameter λ to arrive at another constrained minimization problem of the form (3.7). We continue this process until convergence of the coefficients, meaning no more coefficients smaller than λ in absolute value. The result is (hopefully) a representation of the nonlinear dynamics governing the data X and Y using as few elements of the dictionary as possible. Moreover, [295] proves that the SINDy algorithm approximates solutions of (3.6) while also providing sufficient conditions and rates of convergence. Importantly for computation, [295] proves that the algorithm converges to a local minimizer in a finite number of steps. The general SINDy algorithm is summarized in Algorithm 3.1.

Example 3.1. Let us apply SINDy to data from the Lorenz system

$$\begin{aligned} \dot{x}_1 &= -10x_1 + 10x_2, \\ \dot{x}_2 &= 28x_1 - x_2 - x_1x_3, \\ \dot{x}_3 &= -\frac{8}{3}x_3 + x_1x_2 \end{aligned} \tag{3.8}$$

at the standard chaotic parameter values. We integrate (3.8) to generate synthetic data, resulting in the data matrix

$$X = \begin{bmatrix} x_1(t_1) & x_1(t_2) & x_1(t_3) & \cdots & x_1(t_N) \\ x_2(t_1) & x_2(t_2) & x_2(t_3) & \cdots & x_2(t_N) \\ x_3(t_1) & x_3(t_2) & x_3(t_3) & \cdots & x_3(t_N) \end{bmatrix}, \tag{3.9}$$

with $t_{n+1} = t_n + h$, $t_1 = 0$, $h = 10^{-4}$, and $N = 20000$. The matrix Y will contain estimations of the derivatives of the state variables, written

$$Y = \begin{bmatrix} \delta x_1(t_1) & \delta x_1(t_2) & \delta x_1(t_3) & \cdots & \delta x_1(t_N) \\ \delta x_2(t_1) & \delta x_2(t_2) & \delta x_2(t_3) & \cdots & \delta x_2(t_N) \\ \delta x_3(t_1) & \delta x_3(t_2) & \delta x_3(t_3) & \cdots & \delta x_3(t_N) \end{bmatrix}, \tag{3.10}$$

where

$$\delta x_i(t_n) = \frac{x_i(t_{n+1}) - x_i(t_n)}{h}, \qquad i = 1, 2, 3, \tag{3.11}$$

representing a finite-difference approximation of the time derivative. Such derivative estimations are often necessary with real data, as only the states of the system are available.

We consider a dictionary of monomials in (x_1, x_2, x_3) up to degree 2,

$$\mathcal{D}_2 = \{1, x_1, x_2, x_3, x_1^2, x_1x_2, x_1x_3, x_2^2, x_2x_3, x_3^2\}, \tag{3.12}$$

so that $K = |\mathcal{D}_2| = 10$. Performing system identification using (3.3) and not enforcing sparsity identifies the system from data to be

$$\begin{aligned}
\dot{x}_1 &= -9.9810x_1 + 9.9945x_2 - 0.0005x_1x_3, \\
\dot{x}_2 &= 0.0005 + 27.9882x_1 - 0.9850x_2 - 0.9994x_1x_3, \\
\dot{x}_3 &= -0.0550 + 0.0003x_1 - 0.0003x_2 - 2.6590x_3 + 0.0001x_1^2 \\
&\quad + 0.9984x_1x_2 + 0.0011x_2^2 - 0.0002x_3^2,
\end{aligned} \tag{3.13}$$

where all coefficients smaller than 10^{-4} in absolute value have been set to zero for the presentation. Notice that there are a number of spurious terms in (3.13), many of which can be thresholded to zero by the user to simplify the presentation. However, using SINDy with $\lambda = 0.1$ identifies the system from the same training data to be

$$\begin{aligned}
\dot{x}_1 &= -10.0015x_1 + 10.0000x_2, \\
\dot{x}_2 &= 28.0260x_1 - 1.01229x_2 - 1.0051x_1x_3, \\
\dot{x}_3 &= -2.66715x_3 + 1.0000x_1x_2,
\end{aligned} \tag{3.14}$$

which displays a greater termwise accuracy in the coefficients to the true system (3.8) than would be achieved by simply thresholding out small coefficients in (3.13).

Promoting sparsity with SINDy can also help to mitigate noisy measurements. For illustration we add Gaussian white noise with variance 10^{-4} to measurements. Using (3.3), the identified nonsparse system is

$$\begin{aligned}
\dot{x}_1 &= 0.8790 - 11.1999x_1 + 10.7103x_2 - 0.1192x_3 - 0.0195x_1^2 \\
&\quad + 0.0153x_1x_2 + 0.0315x_1x_3 - 0.0021x_2^2 - 0.0176x_2x_3 + 0.0038x_3^2, \\
\dot{x}_2 &= -0.6066 + 28.7575x_1 - 1.4485x_2 + 0.0822x_3 + 0.0130x_1^2 \\
&\quad - 0.0107x_1x_2 - 1.0193x_1x_3 + 0.0017x_2^2 + 0.0100x_2x_3 - 0.0026x_3^2, \\
\dot{x}_3 &= -2.9297 + 0.0204x_1 - 0.0184x_2 - 2.1012x_3 + 0.0905x_1^2 \\
&\quad + 0.9250x_1x_2 - 0.0005x_1x_3 + 0.0126x_2^2 + 0.0004x_2x_3 - 0.0181x_3^2,
\end{aligned} \tag{3.15}$$

while SINDy with $\lambda = 0.1$ returns the model

$$\begin{aligned}
\dot{x}_1 &= -10.0737x_1 + 10.0556x_2, \\
\dot{x}_2 &= 28.3218x_1 - 1.1329x_2 - 1.0063x_1x_3, \\
\dot{x}_3 &= 0.41845 - 2.68569x_3 + 1.0004x_1x_2.
\end{aligned} \tag{3.16}$$

Although SINDy is closer to the true model (3.8), it still has inaccuracies in the $\dot{x}_3$ equation. These are almost surely due to the finite-difference computations to make Y, whose effect is clearly seen in Figure 3.1.

The previous example shows the utility for the SINDy method to eliminate unnecessary dictionary terms during the model discovery process. However, this process can be corrupted by even small amounts of noise in the measurement data. This failure is primarily attributed to

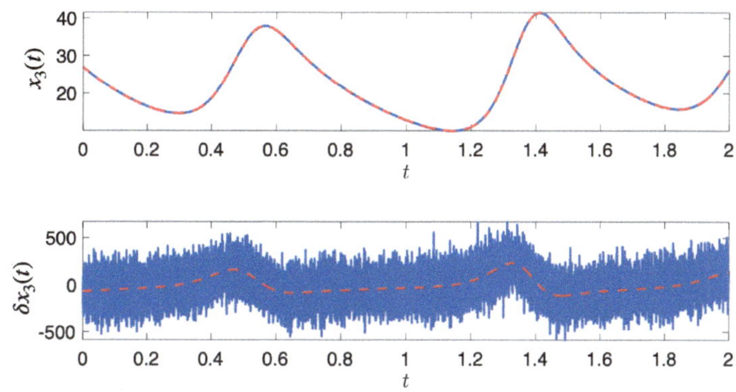

Figure 3.1. *Measurements of the Lorenz system without (dashed red) and with (solid blue) noise. On the top is the $x_3(t)$ component, and on the bottom is its finite-difference approximation of the derivative, $\delta x_3(t)$, as in* (3.11).

the finite-difference approximation of the derivatives. To counter this a suite of model discovery algorithms have been developed that use weak formulations of systems [244, 237, 236, 195, 194, 215]. These methods circumvent approximating derivatives and can be put into the SINDy framework since they again reduce to approximating minimization problems of the form (3.6). Their advantage is that they replace derivatives with integrals which are more robust to noise.

We follow [244] to demonstrate the weak SINDy framework. First, given an ODE

$$\dot{x} = F(x), \tag{3.17}$$

integrating from 0 to t gives the equivalent integral (weak) formulation

$$x(t) = x(0) + \int_0^t F(x(s))\mathrm{d}s. \tag{3.18}$$

Without knowing F, we assume it lies in the span of our dictionary $\mathcal{D}$. Thus, we write

$$x(t) = x(0) + \sum_{k=1}^{K} \Xi_k \int_0^t \theta_k(x(s))\mathrm{d}s \tag{3.19}$$

for each $1 \leq k \leq K$. Assuming further that sequential measurements of the ODE $x(t_n)$ are given with $t_{n+1} = t_n + h$, $t_1 = 0$, $h > 0$ and $1 \leq n \leq N$, we rearrange (3.19) to have

$$x(t_n) - x(t_1) = \sum_{k=1}^{K} \Xi_k \int_0^t \theta_k(x(s))\mathrm{d}s, \qquad n = 1, \ldots, N. \tag{3.20}$$

We estimate the integrals using piecewise constant quadrature over the intervals $[t_n, t_{n+1}]$ to get

$$\int_0^t \theta_k(x(s))\mathrm{d}s \approx \underbrace{h \sum_{j=1}^{n} \theta_k(x(t_j))}_{=:\mathcal{I}_k(t_n)}, \qquad k = 1, \ldots, K. \tag{3.21}$$

With this assumption, the weak SINDy minimization problem is

$$\underset{\Xi \in \mathbb{R}^{M \times K}}{\arg\min} \|\tilde{X} - \Xi \mathcal{I}\|_F^2, \tag{3.22}$$

Algorithm 3.2 Weak-Form SINDy

1. Gather system snapshots X_n
2. Define a dictionary $\mathcal{D} = \{\theta_1, \theta_2, \ldots, \theta_K\}$ and construct the matrices $\tilde{X}$ as in (3.23) and $\mathcal{I}$ as in (3.24)
3. Define a threshold $\lambda > 0$ and perform sequential least squares:
   (a) Obtain the matrix $\Xi = [\xi_{m,k}]_{m,k} = \hat{X}\mathcal{I}^\dagger$
   (b) Obtain minimizer $\tilde{\Xi} = [\tilde{\xi}_{m,k}]_{m,k}$ of $\|\hat{X} - \tilde{\Xi}\mathcal{I}\|_F^2$ subject to $\tilde{\xi}_{m,k} = 0$ if $|\xi_{m,k}| < \lambda$
   (c) Repeat (b) by replacing $\tilde{\Xi}$ with Ξ until these matrices are the same
4. Converged coefficient matrix $\Xi = [\Xi_1 \ldots \Xi_K]$ approximates the ODE $\dot{x} = \Xi_1\theta_1(x) + \Xi_2\theta_2(x) + \cdots + \Xi_K\theta_K(x)$

where

$$\tilde{X} = \begin{bmatrix} | & | & | & \cdots & | \\ 0 & x(t_2) - x(t_1) & x(t_3) - x(t_1) & \cdots & x(t_N) - x(t_1) \\ | & | & | & \cdots & | \end{bmatrix} \in \mathbb{R}^{M\times N} \tag{3.23}$$

and

$$\mathcal{I} = \begin{bmatrix} \mathcal{I}_1(t_1) & \mathcal{I}_1(t_2) & \cdots & \mathcal{I}_1(t_N) \\ \mathcal{I}_2(t_1) & \mathcal{I}_2(t_2) & \cdots & \mathcal{I}_2(t_N) \\ \vdots & \vdots & \ddots & \vdots \\ \mathcal{I}_K(t_1) & \mathcal{I}_K(t_2) & \cdots & \mathcal{I}_K(t_N). \end{bmatrix} \in \mathbb{R}^{K\times N}. \tag{3.24}$$

Sparsity in Ξ for (3.22) is promoted using a sparsity parameter $\lambda > 0$ in the same way as before. This process is summarized in Algorithm 3.2 for the reader.

Example 3.2. Returning to the Lorenz system (3.8) from the previous example, we apply the weak SINDy method to the noisy measurements that produced (3.16). The resulting weak SINDy model is given by

$$\begin{aligned} \dot{x}_1 &= -9.9999x_1 + 10.0001x_2, \\ \dot{x}_2 &= 28.0604x_1 - 1.0187x_2 - 1.0015x_1x_3, \\ \dot{x}_3 &= -2.6668x_3 + 1.0001x_1x_2. \end{aligned} \tag{3.25}$$

Further increasing the variance of the noise to 10^{-3}, 10^{-2}, and 10^{-1} gives similarly accurate discovered models of the Lorenz system (3.8). In all cases the learned coefficients are accurate to within a 5% relative error.

3.2 ▪ Poincaré and Stroboscopic Mappings

The SINDy method is a powerful tool to not only produce new systems from data but also better understand the systems we already have. In this section we demonstrate an application of SINDy to learn nonlinear discrete-time mappings that iterate through coarse-grained solutions of continuous-time systems. These ideas go back to Henri Poincaré in the late nineteenth century with his work on the three-body problem [225]. The three-body problem is now a well-known

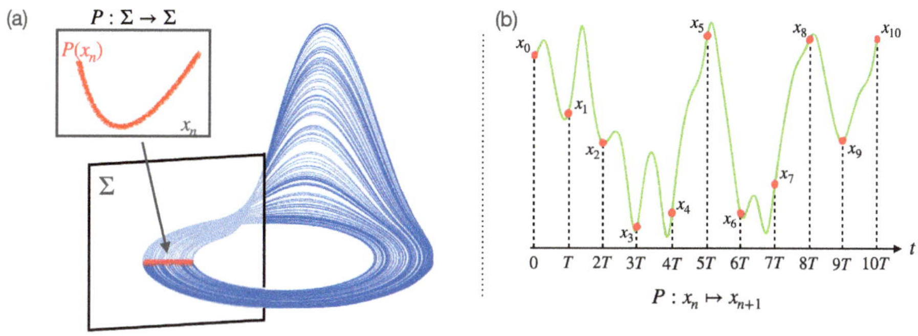

Figure 3.2. *Both Poincaré and stroboscopic maps act to coarse-grain continuous-time dynamics, producing discrete-time iteration schemes.* (a) *A Poincaré section Σ is a plane in phase space that is transverse to the system's flow. Tracking successive intersections with the section leads to a discrete collection of points $\{x_1, x_2, x_3, \dots\}$, and the map P that iterates through these intersections is referred to a Poincaré map.* (b) *To produce a stroboscopic map associated to a nonautonomous periodically forced system, we track the state of the system after each forcing period and determine a map P that iterates through the discrete collection of states.*

example of a chaotic system which has no closed-form solution [106], but Poincaré lacked the current suite of analytical and computational tools to classify it as such. However, Poincaré did notice that although the dynamics of the system were not periodic, they often were **recurrent**, meaning they returned to the same region of phase space over and over. To simplify the output of chaotic systems, Poincaré introduced the use of lower-dimensional transverse subspaces in recurrent regions of phase space to track the intersection of trajectories. The result is a discrete collection of iterates in the subspace for which iterating from one intersection to the next along a trajectory faithfully preserves the chaotic structure of the system.

Poincaré's ideas of moving from a continuous-time output of a system to a discrete collection of points on the trajectory now forms the basis for understanding recurrent systems. For an autonomous system

$$\dot{x} = F(x), \tag{3.26}$$

where $x = x(t) \in \mathbb{R}^d$ and $F : \mathbb{R}^d \to \mathbb{R}^d$, we define a $(d-1)$-dimensional subspace $\Sigma \subset \mathbb{R}^d$, referred to as a **Poincaré section**, and use solutions $x(t)$ of (3.26) to define sequences

$$\{x_n := x(t_n)|\, x(t_n) \in \Sigma,\ t_n > t_{n-1})\}. \tag{3.27}$$

We illustrate the basic process in Figure 3.2. The iterative scheme which takes $x_n \mapsto x_{n+1}$ is referred to as a **Poincaré map**. In parallel, consider a nonautonomous dynamical system

$$\dot{x} = F(x,t), \tag{3.28}$$

where now the function $F : \mathbb{R}^d \times \mathbb{R} \to \mathbb{R}^d$ is periodic in t with period $T > 0$. For solutions $x(t)$ to (3.28) a **stroboscopic map** governs the iterative scheme $x(nT) \mapsto x((n+1)T)$, meaning we only track the solutions after each driving period. Both Poincaré and stroboscopic maps have the advantage that periodic orbits in continuous time reduce to sequences that repeat after some finite number of applications of the map. Thus, these maps can be used to determine the existence of periodic orbits by identifying cyclic points of Poincaré or stroboscopic maps having lower-dimensional phase spaces from the original continuous-time system, (3.26) or (3.28).

In all but a handful of examples it is not possible to determine explicit representations of Poincaré or stroboscopic maps. This means we are often left to investigate simple phenomenological models to explain what is observed in data collected from complex systems. Since

Table 3.1. *Discovered stroboscopic maps $x(n) \mapsto x(n+1)$ for (3.29) using SINDy with different sparsity parameters.*

Sparsity parameter (λ)	SINDy stroboscopic map
0.1	$\mathtt{0.65667}x$
0.05	$\mathtt{0.36489}x - \mathtt{0.09739}$
0.01	$\mathtt{0.36489}x - \mathtt{0.09739}$
0.005	$\mathtt{0.00648}x^5 + \mathtt{0.03060}x^4 - \mathtt{0.02804}x^3$ $+\mathtt{0.36916}x - \mathtt{0.09687}$

phenomenological maps have no direct relationship to specific systems, they cannot be used to forecast or extract dynamics. The SINDy method presents a way of overcoming this gap between theory and experiment by learning maps from data [44, 45]. We demonstrate with the following example while also exploring the importance of the sparsity parameter λ in the discovery process.

Example 3.3. Consider a simple model of an RC circuit [264, Example 8.7.2], given by

$$\dot{x} = \sin(2\pi t) - x, \quad x = x(t) \in \mathbb{R}. \tag{3.29}$$

Since (3.29) is 1-periodic in t, we define the stroboscopic map $x(n) \mapsto x(n+1)$ for $n \in \mathbb{N}$.
Equation (3.29) can be solved using variation of constants, giving the exact solution

$$x(t) = \mathrm{e}^{-t}x_0 + \frac{\sin(2\pi t) - 2\pi\cos(2\pi t) + 2\pi \mathrm{e}^{-t}}{1 + 4\pi^2} \tag{3.30}$$

for an arbitrary initial condition $x_0 \in \mathbb{R}$. Hence, for $n \in \mathbb{N}$ we have

$$x(n+1) = \mathrm{e}^{-1}x(n) + \frac{2\pi(\mathrm{e}^{-1} - 1)}{1 + 4\pi^2} \approx 0.36788x(n) - 0.09812, \tag{3.31}$$

giving explicitly the (linear) stroboscopic map. The map (3.31) has a globally attracting fixed point $x_* = -2\pi/(1 + 4\pi^2)$ corresponding to a globally attracting periodic orbit of (3.29).
We now compare our analytic mapping (3.31) to a SINDy map fit to data collected from the system. To illustrate, we integrate (3.29) with three random initial conditions from $t = 0$ to $t = 200$, sampling each trajectory at integers $n = 0, \dots, 200$. The resulting SINDy maps are presented in Table 3.1 with varying values of the sparsity parameter λ. When λ is chosen too large, SINDy cannot identify the constant term of the map (3.31), while if λ is too small, then unnecessary terms of quadratic order and higher are introduced into the model. This illustrates the importance of the sparsity parameter for identifying SINDy models. In practice tuning λ can be automated by identifying models over a range of λ values and choosing the SINDy models near the Pareto front of the prediction error for the discovered mappings [49, 44, 188].

Around the same time that Poincaré was working on the three-body problem, Gaston Floquet was developing what we now call **Floquet theory** to determine the stability of periodic orbits [88]. Floquet theory starts by linearizing about a periodic orbit, resulting in the ODE

$$\dot{x} = A(t)x, \tag{3.32}$$

where $x = x(t) \in \mathbb{R}^d$ and A is a $d \times d$ matrix, continuous in t such that $A(t+T) = A(t)$ for all t. Here T is the period of the periodic orbit the system was linearized about. Floquet theory gives that the fundamental matrix solution $\Phi(t)$ of (3.32) takes the form [109]

$$\Phi(t) = P(t)\mathrm{e}^{Bt}, \tag{3.33}$$

where P is T-periodic and nonsingular for all t and $B \in \mathbb{R}^{d\times d}$ is a constant matrix.

Floquet theory is a powerful tool to understand the dynamics of a system near a periodic orbit. In particular, the flow of a small initial perturbation from a periodic orbit is approximately governed by the system (3.32). From the fundamental matrix we have the solution $x(t) = \Phi(t)x_0$ to (3.32), and so we define a stroboscopic map that governs the values of the solution after each trip around the periodic orbit, i.e., $x(nT) \mapsto x((n+1)T)$. The fundamental matrix solution gives

$$\begin{aligned} x((n+1)T) &= \Phi((n+1)T)x_0 \\ &= P((n+1)T)\mathrm{e}^{B(n+1)T}x_0 \\ &= \mathrm{e}^{BT}\mathrm{e}^{BnT}x_0 \\ &= \mathrm{e}^{BT}x(nT) \end{aligned} \tag{3.34}$$

since $\Phi(0) = P(0)$ is the identity mapping by definition of the fundamental matrix solution, while periodicity implies $P(nT) = P(0)$ for all $n \in \mathbb{N}$. Thus, the resulting stroboscopic map is linear with the stability of the periodic orbit captured by the eigenvalues of e^{BT}.

The downside is that Floquet theory can only be used to capture *local* stability properties of periodic orbits in nonlinear systems. That is, the linear stroboscopic map (3.34) only approximates the dynamics in a small tubular neighborhood of the periodic orbit, while it cannot help to understand the dynamics of solutions that start outside of this neighborhood. To capture the global behavior of the system, including transient dynamics and multiple basins of attraction, we consider a Poincaré map that is orthogonal to the periodic orbit at a point. This makes the periodic orbit a fixed point of the associated Poincaré map, while the global dynamics in the section can capture the nonlinear approach to the orbit away from the region of Floquet theory's validity. Thus, Poincaré maps provide a form of *nonlinear Floquet theory* while also providing utility for studying the existence and stability of other invariant manifolds such as tori and chaotic attractors. So, using SINDy to discovering Poincaré maps means that we can develop a nonlinear Floquet theory to study nonlinear systems.

Example 3.4. Let us consider the planar system

$$\begin{aligned} \dot{x}_1 &= x_1 - \omega x_2 - x_1(x_1^2 + x_2^2), \\ \dot{x}_2 &= \omega x_1 + x_2 - x_2(x_1^2 + x_2^2), \end{aligned} \tag{3.35}$$

where $\omega > 0$ is fixed. System (3.35) has a unique limit cycle given by the unit circle, written

$$\gamma(t) := \begin{bmatrix} \cos(\omega t) \\ \sin(\omega t) \end{bmatrix}. \tag{3.36}$$

Linearizing (3.35) about the periodic orbit $\gamma(t)$ leads to the time-periodic ODE

$$\begin{bmatrix} \dot{x}_1 \\ \dot{x}_2 \end{bmatrix} = \begin{bmatrix} -2\cos^2(\omega t) & -\omega - 2\cos(\omega t)\sin(\omega t) \\ \omega - 2\cos(\omega t)\sin(\omega t) & -2\sin^2(\omega t) \end{bmatrix} \begin{bmatrix} x_1 \\ x_2 \end{bmatrix} \tag{3.37}$$

of the form (3.32). One solution of (3.37) is $(x_1(t), x_2(t)) = \dot{\gamma}$, the derivative of the periodic orbit. This property is universal to autonomous systems [109] and corresponds to the translational invariance along the periodic orbit; i.e., $\gamma(t+\alpha)$ is also a periodic orbit of (3.35) for all $\alpha \in \mathbb{R}$.

To find another linearly independent solution of (3.37), we transform (3.35) to polar coordinates by setting $x_1 = r\cos(\theta)$ and $x_2 = r\sin(\theta)$. The result is

$$\begin{aligned} \dot{r} &= r - r^3, \\ \dot{\theta} &= \omega, \end{aligned} \tag{3.38}$$

for which the nontrivial steady state $r = 1$ of the $\dot{r}$ equation is exactly the limit cycle γ. So, linearizing the $\dot{r}$ equation about this steady state gives

$$\dot{r} = -2r, \tag{3.39}$$

which has solution $r(t) = r_0 e^{-2t}$ for any $r_0 > 0$. The $\dot{\theta}$ equation in (3.38) completely decouples from the radial components, resulting in the solution $\theta(t) = \omega t + \theta_0$. Since the periodic orbit $\gamma(t)$ starts with initial phase 0 in the plane, we take $\theta_0 = 0$ to enforce that our solution follows the linearization about the periodic orbit in (3.37). Moving back to Euclidean coordinates we have the second solution of (3.37), given by

$$p(t) = r_0 e^{-2t} \begin{bmatrix} \cos(\omega t) \\ \sin(\omega t) \end{bmatrix}. \tag{3.40}$$

With these two solutions, we arrive at the fundamental matrix solution of (3.37),

$$\Phi(t) = \begin{bmatrix} e^{-2t}\cos(\omega t) & -\sin(\omega t) \\ e^{-2t}\sin(\omega t) & \cos(\omega t) \end{bmatrix}, \tag{3.41}$$

whose columns are composed of the orthogonal vectors $p(t)$ and $\omega^{-1}\dot{\gamma}(t)$, respectively. The period of $\gamma(t)$ is $2\pi/\omega$, and so the linear stroboscopic map (3.34) is governed by the matrix

$$\Phi(2\pi/\omega) = \begin{bmatrix} e^{-4\pi/\omega} & 0 \\ 0 & 1 \end{bmatrix}. \tag{3.42}$$

Eigenvalues of this matrix are 1, representing the neutral direction of translations along the periodic orbit, and $e^{-4\pi/\omega} < 1$, giving that the periodic orbit is locally asymptotically stable.

Since (3.35) is nonlinear, the matrix (3.42) can only capture the dynamics of the system near the periodic orbit γ. Notice that the origin is a steady-state solution of (3.35) and that this state is not captured by the linear stroboscopic map coming from Floquet theory. Therefore, Floquet theory has no way of capturing two distinct solutions of a given system and so cannot describe the approach from one to the other. To overcome this we seek to identify a Poincaré map.

We will take the nonnegative horizontal axis, $\Sigma = \{(x_1, 0) \in \mathbb{R}^2 |\ x_1 \geq 0\}$, as our Poincaré section. This Poincaré section is equivalently defined in polar coordinates by the condition $\theta(t) \equiv 0 \mod 2\pi$, allowing for the explicit determination of the Poincaré map. Take $(r(0), \theta(0)) = (r_0, 0)$ to be an initial condition in the Poincaré section. Then $\theta(t) = \omega t$, and the first time $t_* > 0$ that the solution hits the Poincaré section again is $t_* = 2\pi/\omega$. Using separation of variables we have

$$\begin{aligned} &\dot{r} = r - r^3 \\ \implies &\int \frac{dr}{r(1-r^2)} = \int dt \\ \implies &\ln(r) - \frac{1}{2}\ln(1-r^2) = t + C \\ \implies &r(t) = \sqrt{\frac{1}{1 + (r_0^{-2} - 1)e^{-2t}}}, \end{aligned} \tag{3.43}$$

where the final step uses $r(0) = r_0$ to solve for the constant of integration C. The solution hits the Poincaré section again at $t = t_*$, resulting in the Poincaré map, $P : \Sigma \to \Sigma$, given by

$$P(x_1) = \sqrt{\frac{1}{1 + (x_1^{-2} - 1)\mathrm{e}^{-4\pi/\omega}}} \tag{3.44}$$

for each $x_1 \geq 0$. The evaluation at $x_1 = 0$ is considered in the limit $x_1 \to 0^+$ so that $P(0) = 0$.

There are two things that we should notice here in the Poincaré map. First, the linearization about the fixed point $x_1 = 1$, $P'(1)$, is exactly $\mathrm{e}^{-4\pi/\omega}$. Thus, the linearization of the Poincaré map about the intersection with the limit cycle $\gamma(t)$ in (3.35) corresponds exactly to the eigenvalue of (3.42) that demonstrates stability of the orbit. The second eigenvalue of (3.42) at 1 is not present in the Poincaré map linearization, as this eigenvalue represents the transversality of the section. Second, the map P is not a linear combination of elementary functions (polynomials, sinusoids, etc.) and so it is unlikely for SINDy to discover exactly with a user-defined dictionary. Nonetheless, we demonstrate that SINDy can qualitatively capture the dynamics of the map.

Our numerical experiments will set $\omega = 10\pi$ so that convergence to the periodic orbit is slow enough to collect sufficient training data to observe transient approaches to it. Training data is collected from 10 trajectories, of which 8 initial conditions are randomized in the Poincaré section with $x \in [0, 5]$, one initial condition with $x = 0$ to provide knowledge of the unstable origin, and one initial condition with $x = 0.0001$ to demonstrate the passage from the unstable fixed point at $x = 0$ to the stable fixed point at $x = 1$. We consider a dictionary of monomials in x up to degree 5.

With a sparsity parameter of $\lambda = 0.1$, SINDy provides the approximate Poincaré mapping

$$\tilde{P}_{0.1}(x_1) = \mathtt{1.1348}x_1 - \mathtt{0.13672}x_1^3. \tag{3.45}$$

The map $\tilde{P}_{0.1}$ has a fixed point at $x_* = 0.99300$, with $\tilde{P}'_{0.1}(0.99300) = 0.73040$, differing on the order of the sparsity parameter from the true value $P'(1) \approx 0.67032$. Decreasing the sparsity parameter to $\lambda = 0.01$ results in the nonsparse quintic SINDy map

$$\tilde{P}_{0.01}(x_1) = \mathtt{1.2229}x_1 + \mathtt{0.02357}x_1^2 - \mathtt{0.44146}x_1^3 + \mathtt{0.23788}x_1^4 - \mathtt{0.045545}x_1^5. \tag{3.46}$$

This map now has the fixed point $x_* = 0.99190$, with stability $\tilde{P}'_{0.01}(0.99190) = 0.67479$, agreeing with $P'(1)$ to two decimal places. Both SINDy models approximately capture the location of the fixed point $x_1 = 1$ in the Poincaré section and its stability. Moreover, they both have the second unstable fixed point at $x_1 = 0$, although their linearization about this fixed point differs significantly from the linearization $P'(0) = 1$ in the true Poincaré map. This means that the global section dynamics of being pushed away from $x_1 = 0$ and into $x_1 = 1$ is not completely described by the SINDy models. However, they do provide an accurate approximation of this process. One can fine-tune these maps by including higher-order monomials in the dictionary at the cost of obtaining models that are not sparse. For example, using a dictionary of monomials up to degree 20 results in a linearization about $x_1 = 0$ of 1.01157, but the model uses all monomials and contains coefficients on the order of 10^4. Having no knowledge of which terms to include in the dictionary is certainly a shortcoming of both EDMD and SINDy, although approximation theorems, such as the Stone–Weierstrass theorem [224], can be used to guide these choices.

3.3 ▪ Stabilizing Periodic Orbits

A great deal of time in studying dynamical systems is spent determining the stability of solutions. Moreover, it is the stable solutions that garner the most attention. The explanation for this is simple: Stable solutions are locally attracting, and so they form the observable dynamics of the system. So long as we start close to a stable solution, we will converge into it and be stuck there for the rest of time. On the other hand, unstable solutions push nearby trajectories away and therefore cannot be observed in the system without initializing exactly on the unstable solution or its stable manifold. Numerical error makes this practically impossible, and so unstable solutions are seldom seen in simulation. Despite this, unstable trajectories play a critical role by separating out regions of phase space and guiding trajectories for finite time.

Ott, Grebogi, and Yorke [210] proposed a method of introducing slight manipulations into a system to reorient trajectories back to unstable solutions. Their work focuses on chaotic systems, where one can find infinitely many unstable periodic orbits densely embedding inside of a chaotic attractor. However, their methods apply broadly to unstable orbits in nonchaotic systems as well. They propose kicking the system each time a trajectory pierces a Poincaré section, with the appropriate kick determined by the associated Poincaré map. The precise nature of these perturbations requires knowing the Poincaré map, which from the discussion in the previous section is no easy task. Fortunately, data-driven discovery methods such as SINDy are able to obtain good approximations of Poincaré maps using only data gathered from the system, thus allowing for the widespread applicability of this method of stabilizing periodic orbits.

To begin, we consider a parameter-dependent discrete dynamical system

$$x_{n+1} = F(x_n, \mu), \tag{3.47}$$

where $F : \mathbb{R}^d \times \mathbb{R}^p \to \mathbb{R}^d$ is smooth. Our first example in this section will illustrate the stabilization method using a known discrete-time mapping, and then we will combine stabilizing with data-driven discovery of a Poincaré map using SINDy to apply the method to a continuous-time system. Suppose that at some parameter value $\mu = \bar{\mu}$ the map (3.47) has a fixed point, $\bar{x}$, i.e., $\bar{x} = F(\bar{x}, \bar{\mu})$. Stability of this fixed point comes from the eigenvalues of the matrix $A := F_x(\bar{x}, \bar{\mu})$, the linearization of (3.47) about $(x, \mu) = (\bar{x}, \bar{\mu})$, for which we assume that A has eigenvalues outside the unit circle in the complex plane, meaning $\bar{x}$ is unstable.

The idea put forth in [210] is to introduce small, precise parameter perturbations into (3.47) at each iteration to keep trajectories close to the unstable fixed point $\bar{x}$. Following [239], this can be achieved by inducing the parameter manipulation $\mu = \mu_n$ at each iterate by

$$\mu_n = \begin{cases} \bar{\mu} + C(x_n - \bar{x}), & \|x_n - \bar{x}\| \le \eta, \\ \bar{\mu}, & \|x_n - \bar{x}\| > \eta, \end{cases} \tag{3.48}$$

for a yet-to-be-specified controller matrix $C \in \mathbb{R}^{p \times d}$ and a user-defined threshold $\eta > 0$. The idea is that when iterates of (3.47) are close to the unstable fixed point $\bar{x}$, as measured by $\|x_n - x\| \le \eta$, we attempt to choose C in such a way that they remain close, while if the iterates are far away from $\bar{x}$, we let the system run without intervention. The reason that [210] focuses on chaotic systems is that ergodicity along the chaotic attractor guarantees that if $\bar{x}$ is in the attractor, eventually almost every initial condition must come into an η-neighborhood of $\bar{x}$, and so the controller will eventually be turned on to keep the trajectory in this neighborhood.

The question now becomes, how do we choose C? Linearizing (3.47) about $(x, \mu) = (\bar{x}, \bar{\mu})$ with μ as in (3.48), we have

$$x_{n+1} \approx \bar{x} + A(x_n - x) + B(\mu - \mu_n) = \bar{x} + (A + BC)(x_n - x), \tag{3.49}$$

where again $A = F_x(\bar{x}, \bar{\mu})$ and now $B = F_\mu(\bar{x}, \bar{\mu})$. Trajectories of (3.47) will converge to $\bar{x}$ if C is chosen so that $A + BC$ is stable; i.e., all of its eigenvalues lie inside the unit circle. Finding a C that makes $A + BC$ stable is referred to as the **pole-placement method**, which is a standard in control theory [75] that can be automated using semidefinite programming.

We first use the fact that a matrix M is stable if and only if there exists $P \succ 0$ such that

$$P - M^{\intercal} P M \succ 0. \tag{3.50}$$

Here the subscript $\intercal$ denotes the matrix transpose, and $\succ 0$ means that the matrix is positive definite; i.e., it is symmetric with all eigenvalues positive. Stability via (3.50) comes from the existence of a discrete-time Lyapunov function $V(x) = x^{\intercal} P x$ to prove the stability of $x = 0$ in the linear system $x_{n+1} = Mx_n$ (see Example 4.1 for more details). Using (3.50), we aim to find C and $P \succ 0$ so that

$$P - (A+BC)^{\intercal} P (A+BC) \succ 0. \tag{3.51}$$

Unfortunately, the above equation is nonlinear in the unknowns P and C, making it difficult to solve. Fortunately, [30] characterizes an equivalent formulation that is linear in the unknowns.

First, we use the Schur complement to equivalently write (3.51) as

$$\begin{bmatrix} P & (A+BC)^{\intercal} P \\ P(A+BC) & P \end{bmatrix} \succ 0. \tag{3.52}$$

Now, since we require $P \succ 0$, any P that is identified must be invertible. Letting $Q = P^{-1}$ and multiplying (3.52) on the left and right by the block diagonal matrix $\mathrm{diag}(Q, Q)$ gives

$$\begin{bmatrix} Q & Q(A+BC)^{\intercal} \\ (A+BC)Q & Q \end{bmatrix} \succ 0. \tag{3.53}$$

Since Q is symmetric we have

$$Q(A+BC)^{\intercal} = QA^{\intercal} + QC^{\intercal}B^{\intercal} = QA^{\intercal} + Q^{\intercal}C^{\intercal}B^{\intercal} = QA^{\intercal} + (CQ)^{\intercal}B^{\intercal} \tag{3.54}$$

and $(A+BC)Q = AQ + BCQ$, meaning that (3.53) now only contains unknown terms of the form Q and CQ. Setting $Y = CQ$ we have

$$\begin{bmatrix} Q & QA^{\intercal} + Y^{\intercal}B^{\intercal} \\ AQ + BY & Q \end{bmatrix} \succ 0, \tag{3.55}$$

which is now linear in the unknowns $(Q, Y) \in \mathbb{R}^{d\times d} \times \mathbb{R}^{p\times d}$. The inequality (3.55) constitutes a **linear matrix inequality** (LMI) and can be solved using semidefinite programming. Numerically, strict inequalities cannot be handled, so we introduce a small $\varepsilon > 0$ and solve the LMI

$$\begin{bmatrix} Q - \varepsilon I_d & QA^{\intercal} + Y^{\intercal}B^{\intercal} \\ AQ + BY & Q - \varepsilon I_d \end{bmatrix} \succeq 0, \tag{3.56}$$

where I_d is the $d \times d$ identity matrix. The reason for this is that $\succ$ functions in the same way as usual inequalities, and therefore (3.56) implies

$$\begin{bmatrix} Q & QA^{\intercal} + Y^{\intercal}B^{\intercal} \\ AQ + BY & Q \end{bmatrix} \succeq \varepsilon \mathrm{diag}(I_d, I_d) \succ 0, \tag{3.57}$$

thus verifying (3.55). In practice we will fix $\varepsilon = 10^{-3}$.

Example 3.5. We illustrate this fixed point stabilization procedure on the de Jong map [222]

$$\begin{bmatrix} x_1 \\ x_2 \end{bmatrix}_{n+1} = \begin{bmatrix} \sin(ax_2) - \cos(bx_1) \\ \sin(cx_1) - \cos(dx_2) \end{bmatrix}_n. \tag{3.58}$$

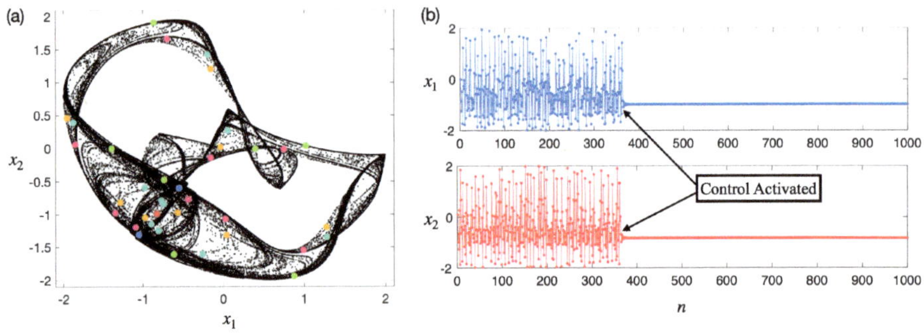

Figure 3.3. (a) *The attractor (black) of the de Jong map* (3.58) *with an array of unstable cyclic points: (•) period* 1 *(fixed point), (•) period* 2, *(•) period* 7, *(•) period* 8, *(•) period* 9, *and (•) period* 10. (b) *A controlled trajectory that is forced to stay near the unstable fixed point.*

We fix $(b, c, d) = (-1.899, 1.381, -1.506)$ and consider a to be the control parameter. In particular, at $a = 0.970$ the map is chaotic, with the attractor plotted in Figure 3.3(a). There is a unique fixed point inside of the attractor, plotted in Figure 3.3 as a red dot (•) and given by

$$(\bar{x}_1, \bar{x}_2) = (-0.8243, -0.9891). \tag{3.59}$$

The resulting linearized matrices A and B therefore become

$$A = \begin{bmatrix} -1.8990 & 0.5568 \\ 0.5788 & -1.5010 \end{bmatrix}, \qquad B = \begin{bmatrix} -0.5677 \\ 0 \end{bmatrix}, \tag{3.60}$$

with A having eigenvalues -2.3015 and -1.0985, so the fixed point is unstable.

To identify the control matrix C using (3.56) we use YALMIP [177] to translate the problem into a semidefinite program and MOSEK [15] to solve it. The resulting control matrix is

$$C = \begin{bmatrix} -4.5101 & 4.1923 \end{bmatrix}. \tag{3.61}$$

Eigenvalues of $A + BC$ are -0.0822 and -0.7573, both residing inside the unit circle, thus verifying that we have indeed stabilized the fixed point (3.59). Figure 3.3 shows controlled iterations of the de Jong map using $\eta^2 = 0.05$. The initial transient is from the trajectory not coming close enough to the fixed point to initiate the control (3.48). After approximately 360 iterates the trajectory is within the η-threshold of the fixed point, and the controller is activated. With the controller activated, the trajectory remains close to the fixed point for all further iterates.

If the map (3.47) is taken to be a Poincaré map, then a fixed point corresponds to a periodic orbit that pierces the Poincaré section at a unique point. It is also possible to have periodic orbits that intersect the Poincaré section at multiple points before completing a full period. In the mapping (3.47) these points are cyclic, meaning after a finite number of iterates the sequence goes back to where it started. Precisely, the set $\{\bar{x}_1, \bar{x}_2, \dots, \bar{x}_k\} \subset \mathbb{R}^d$, for some $k \geq 1$, is said to be a **period *k* orbit**, or ***k*-cycle**, of the map (3.47) at $\mu = \bar{\mu}$ if

$$\bar{x}_2 = F(\bar{x}_1, \bar{\mu}), \quad \bar{x}_3 = F(\bar{x}_2, \bar{\mu}), \quad \dots \quad \bar{x}_1 = F(\bar{x}_k, \bar{\mu}). \tag{3.62}$$

Notice that each $\bar{x}_j$ in the sequence is a fixed point of F^k, the kth iterate mapping of (3.47). Therefore, the stability of a period k orbit is determined by linearizing around any of the $\bar{x}_j$ as a fixed point of F^k. The chain rule gives that

$$F_x^k(\bar{x}_k, \bar{\mu}) = A_k A_{k-1} \cdots A_2 A_1, \tag{3.63}$$

where $A_j := F_x(\bar{x}_j, \bar{\mu})$. If at least one eigenvalue of $F_x^k(\bar{x}_k, \bar{\mu})$ has modulus larger than 1, then the period k orbit is unstable.

To stabilize a period k orbit, we could implement the above stabilization procedure on the kth iterate map, F^k. However, if k is large, this means that there is a long time between parameter perturbations to stabilize the orbit. This would give the trajectory a lot of time to escape a neighborhood of the period k orbit points, necessitating larger parameter kicks to drive the trajectory back on course. Alternatively, we can apply a parameter kick at each iteration of the map F when trajectories become close enough to *any* of the elements of the period k orbit. This helps to keep the trajectory on course and stabilize the period k orbit with smaller parameter kicks. Following [45], we produce LMIs that identify the kicks to stabilize k-cycles.

We introduce the parameter perturbations

$$\mu_n = \begin{cases} \bar{\mu} + C_j(x_n - \bar{x}_j), & \|x_n - \bar{x}_j\| \leq \eta,\ j = 1, \ldots, k, \\ \bar{\mu}, & \text{otherwise}, \end{cases} \tag{3.64}$$

which depend on the proximity of the iterates to each of the elements of the period k orbit. In the neighborhood of $(x, \mu) = (\bar{x}_k, \bar{\mu})$ the linearization of F^k becomes

$$x_{n+k} \approx \bar{x}_k + (A_k + B_k C_k)(A_{k-1} + B_{k-1}C_{k-1}) \cdots (A_1 + B_1 C_1)(x_n - \bar{x}_k), \tag{3.65}$$

where A_j is as above and $B_j = F_\mu(\bar{x}_j, \bar{\mu})$. Therefore, we require $C_j \in \mathbb{R}^{p \times d}$ so that the matrix

$$(A_k + B_k C_k)(A_{k-1} + B_{k-1}C_{k-1}) \cdots (A_1 + B_1 C_1) \tag{3.66}$$

is stable. Note that this is not equivalent to simply having each $A_j + B_j C_j$ be stable, and so we cannot just stabilize each element of the k-cycle individually.

Instead, to find $(C_1, C_2, \ldots, C_k)$ that make (3.66) stable, we iteratively identify them using LMIs, starting from $j = k$. Start by obtaining a matrix C_k so that $D_k := A_k + B_k C_k$ is stable using (3.56). Then, with this C_k we identify C_{k-1} so that the matrix

$$\begin{aligned} D_{k-1} &:= D_k(A_{k-1} + B_{k-1}C_{k-1}) \\ &= D_k A_{k-1} + D_k B_{k-1} C_{k-1} \end{aligned} \tag{3.67}$$

is stable. To do this we simply replace A and B in (3.56) with $D_k A_{k-1}$ and $D_k B_{k-1}$, respectively. Continuing this process, we iteratively define

$$\begin{aligned} D_j &:= D_k D_{k-1} \cdots D_{j+1}(A_j + B_j C_j) \\ &= D_k D_{k-1} \cdots D_{j+1} A_j + D_k D_{k-1} \cdots D_{j+1} B_j C_j \end{aligned} \tag{3.68}$$

and find C_j so that D_j is stable for each $j = 1, \ldots, k$. Notice that the final matrix, D_1, is exactly (3.66) and therefore gives the desired stability for the k-cycle.

Example 3.6. Recall that Figure 3.3(a) plotted a variety of periodic points embedded in the chaotic attractor of the de Jong map (3.58), any of which can be stabilized with these methods. For example, there is a period 2 orbit

$$(\bar{x}_1^{(1)}, \bar{x}_2^{(1)}) = (-1.0466, -1.3133), \qquad (\bar{x}_1^{(2)}, \bar{x}_2^{(2)}) = (-0.5515, -0.5963), \tag{3.69}$$

represented by the blue dots (•) in Figure 3.3(a). Following as above returns the control matrices

$$C_1 = \begin{bmatrix} -5.0546 & 5.0229 \end{bmatrix}, \qquad C_2 = \begin{bmatrix} -4.4441 & 3.0368 \end{bmatrix}, \tag{3.70}$$

which results in the matrix $(A_2 + B_2 C_2)(A_1 + B_1 C_1)$ having eigenvalues 0 and -0.0199.

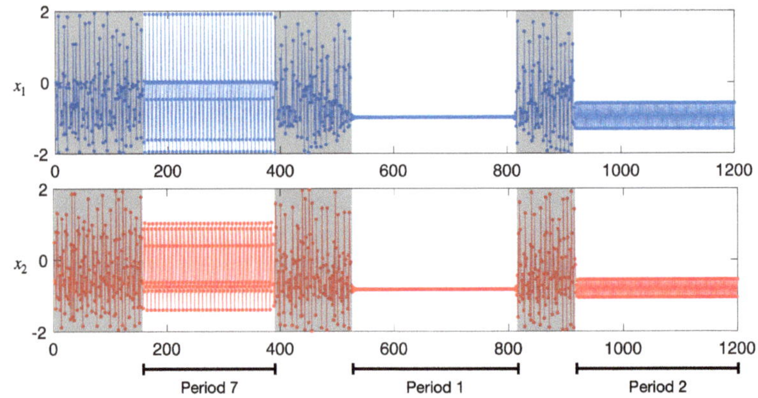

Figure 3.4. *Controlled switching orbits of the de Jong map* (3.58). *The trajectory is first forced to stay near a period* 7 *orbit for* 250 *iterates, then near the fixed point (period* 1 *orbit) for* 250 *iterates, and then finally to a period* 2 *orbit indefinitely. Shaded regions represent transitory states when the control is not activated because the trajectory is not close enough to the desired periodic orbit to stabilize.*

We can implement this stabilization procedure for k-cycles of any length while further designing controllers that switch between stabilizing different cycles. Figure 3.4 demonstrates this with a trajectory of the de Jong map that is initially driven to a period 7 orbit, where it is held for 250 iterates before the controller is turned off again. After this the controller is switched to stabilizing the fixed point from Example 3.5, again remaining close for 250 iterates of the map before being turned off. Finally, the controller switches to stabilizing the period 2 orbit (3.69) using the control matrices (3.70) for the remainder of the iterates of the map.

This stabilization framework can be combined with Poincaré map discovery to control continuous-time systems. That is, we apply a small, precise parameter perturbation each time a trajectory of an ODE hits the Poincaré section to stabilize an unstable periodic orbit. We can implement the SINDy method to learn the associated (parameter-dependent) Poincaré map of the form (3.47) using only data from the system and then identify the system's unstable periodic orbits by extracting its cyclic points. Data is gathered from the Poincaré section over a range of parameter values that includes the focal value, $\mu = \bar{\mu}$, resulting in data matrices $X_\mu \in \mathbb{R}^{(d+p)\times N}$ and $Y \in \mathbb{R}^{d\times N}$ with $N \geq 1$ data points. We augment the state variable, denoted X_μ, so that its first d rows are the Poincaré section variables and the last p are the parameter value at which the data is gathered. A SINDy dictionary thus requires functions $\theta : \mathbb{R}^{d+p} \to \mathbb{R}^d$ acting as $(x,\mu) \mapsto \theta(x,\mu)$ to identify a parameter-dependent Poincaré mapping.

Example 3.7. To implement periodic orbit stabilization in continuous time, we consider Sprott's chaotic jerk model [259]

$$\frac{\mathrm{d}^3x}{\mathrm{d}t^3} + \mu\frac{\mathrm{d}^2x}{\mathrm{d}t^2} - \left(\frac{\mathrm{d}x}{\mathrm{d}t}\right)^2 + x = 0, \qquad \mu \in \mathbb{R}. \tag{3.71}$$

Equation (3.71) can be equivalently cast as the first-order system

$$\begin{aligned} \dot{x}_1 &= x_2, \\ \dot{x}_2 &= x_3, \\ \dot{x}_3 &= -x_1 - \mu x_3 + x_2^2 \end{aligned} \tag{3.72}$$

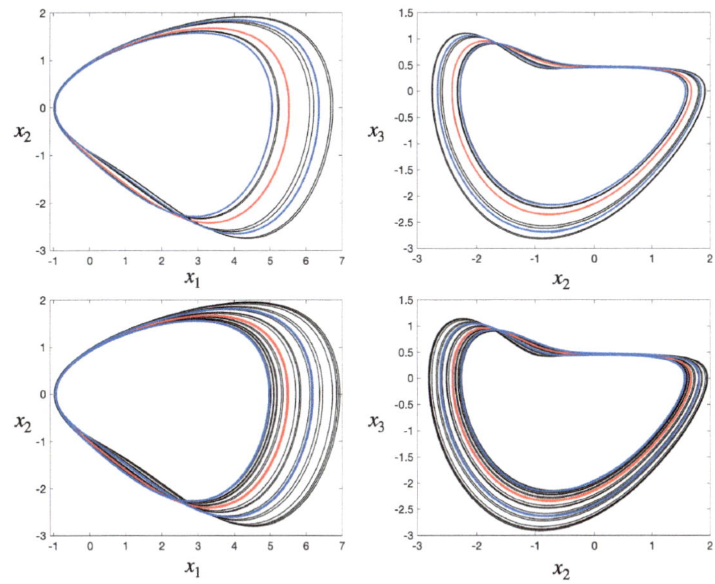

Figure 3.5. *Stabilized period 1 (red) and period 2 (blue) orbits of (3.72) in the (x_1, x_2)- and (x_2, x_3)-planes. Top panels use $\bar{\mu} = 2.06$, where uncontrolled trajectories fall into a period 8 attractor (black). Bottom panels use $\bar{\mu} = 2.05$, where uncontrolled trajectories fall into a chaotic attractor (black).*

upon introducing $x_2 = \mathrm{d}x/\mathrm{d}t$ and $x_3 = \mathrm{d}^2x/\mathrm{d}t^2$. When the parameter μ is decreased from approximately 2.15 to 2, system (3.72) undergoes a sequence of period-doubling bifurcations leading to chaos. In this example we stabilize periodic orbits of (3.71) for different values of μ.

We begin by taking $\bar{\mu} = 2.06$, at which point system (3.72) has a periodic attractor and (at least) three unstable periodic orbits nearby. The Poincaré section is taken to be $x_2 = 0$ with crossing condition $\dot{x}_2 < 0$. In this Poincaré section the periodic attractor is a period 8 orbit, while the three unstable periodic orbits are periods 1, 2, and 4. To gather training data, we gather data in the Poincaré section from integrating (3.72) with μ ranging from 2.04 to 2.08 in increments of 0.001. SINDy is applied with a dictionary of cubic monomials in (x_1, x_3, μ) and sparsity parameter $\lambda = 0.01$ to learn a μ-dependent Poincaré map.

At $\mu = 2.06$ the discovered Poincaré map has a fixed point given by

$$(\bar{x}_1, \bar{x}_3) = (\texttt{5.5278}, \texttt{-2.1349}), \tag{3.73}$$

corresponding to the intersection of an unstable periodic orbit with the Poincaré section $x_2 = 0$. It is interesting to note that there is no value of $\mu \in [2.04, 2.08]$ for which a fixed point in the Poincaré section is stable, meaning that our training data contains no explicit trace of this fixed point. Nonetheless, the discovered mapping is able to infer the existence of the fixed point at $\mu = 2.06$, for which linearizing the discovered Poincaré map about this results in a matrix with eigenvalues `-1.6498` and `-0.0844`, reinforcing that it is indeed unstable. From the discovered Poincaré map we identify the control matrix

$$C = \begin{bmatrix} \texttt{0.1779} & \texttt{0.1347} \end{bmatrix}, \tag{3.74}$$

for which the stabilized matrix $A + BC$, as in (3.49), has eigenvalues `0.0000` and `0.0226`. In Figure 3.5 we present the stabilized period 1 orbit against the period 8 attractor.

The discovered Poincaré map also identifies an unstable period 2 orbit at $\mu = 2.06$, given by

$$(\bar{x}_1^{(1)}, \bar{x}_3^{(1)}) = (\texttt{5.1487}, \texttt{-2.0008}), \qquad (\bar{x}_1^{(2)}, \bar{x}_3^{(2)}) = (\texttt{6.4246}, \texttt{-2.4485}). \tag{3.75}$$

Identifying the control matrices using linearizations about these points in the discovered Poincaré map allows for the stabilization of a period 2 orbit in (3.72) as well. The result is presented in Figure 3.5, along with the period 1 orbit described previously, both using $\eta^2 = 0.05$. There is also an unstable period 4 orbit that has been stabilized using these methods, but it is difficult to differentiate from the attracting period 8 orbit and so is omitted from Figure 3.5 for clarity.

Since the discovered Poincaré mapping is trained over the parameter range $\mu \in [2.04, 2.08]$, we can use it to stabilize periodic orbits at parameter values where the dynamics are more complex. For example, at $\mu = 2.05$ the attractor of the Sprott system (3.72) is chaotic. We repeat the above to stabilize the period 1 and 2 orbits at $\mu = 2.05$, with the results plotted in Figure 3.5.

3.4 ▪ Data-Driven Averaging

In Section 2.2 we were introduced to multiscale signals exhibiting periodic dynamics on short timescales and a slow drift on longer ones. Mathematically this timescale separation of fast oscillations and slow drifts can be observed in nonautonomous systems of the form

$$\dot{x} = \varepsilon f(x, t, \varepsilon), \qquad x(0) = x_0, \tag{3.76}$$

where $0 < \varepsilon \ll 1$ is the timescale separation parameter. When the right-hand side f of (3.76) is periodic in t with period $T > 0$, the theory of averaging [109, 243, 203] gives that solutions $x(t)$ to (3.76) are well approximated by solutions $y(t)$ of the autonomous system

$$\dot{y} = \frac{\varepsilon}{T} \int_0^T f(y, s, 0)\mathrm{d}s =: \varepsilon \bar{f}(y), \qquad y(0) = x_0. \tag{3.77}$$

Precisely, there exists a $C, L > 0$ such that $|x(t) - y(t)| \leq C\varepsilon$ for all $0 \leq t \leq L/\varepsilon$.

What averaging theory shows is that the periodicity in t of (3.76) leads to fast-scale, small-amplitude oscillations in $x(t)$ of size $\mathcal{O}(\varepsilon)$, while (3.77) averages out these fast oscillations, leaving only the slow-scale drift behind them that emerges on long timescales, i.e., on the order of $1/\varepsilon$. Roughly, one should think of the signal $x(t)$ having the decomposition $x(t) = p(t) + y(t)$, where $y(t)$ is the solution to the averaged equation (3.77) and $p(t)$ is a T-periodic signal with $|p(t)| \leq C\varepsilon$. Figure 3.6 provides an illustration of an application of averaging, and the following example seeks to demonstrate the theory analytically.

Example 3.8. Let us consider the nonautonomous logistic equation

$$\dot{x} = \varepsilon x(1 - x + \sin(2\pi t)), \qquad x(0) = x_0 \in \mathbb{R}_+, \tag{3.78}$$

which is amenable to averaging theory when $\varepsilon > 0$ is small. Averaging the right-hand side of (3.78) over a forcing period gives the autonomous averaged system

$$\dot{y} = \varepsilon y(1 - y), \qquad y(0) = x_0, \tag{3.79}$$

the standard **logistic equation**. Using separation of variables, the solution of (3.79) is

$$y(t) = \frac{x_0}{x_0 + (1 - x_0)\mathrm{e}^{-\varepsilon t}}, \tag{3.80}$$

where we note the dependence on the "slow time" variable εt. Moreover, $y(t) \to 1$ as $t \to \infty$ for all $x_0 > 0$. Since the averaged equation settles into a steady state, the averaged solution is valid for all $t \geq 0$, and $x(t)$ settles into periodic motion about $x = 1$ with $\mathcal{O}(\varepsilon)$ amplitude.

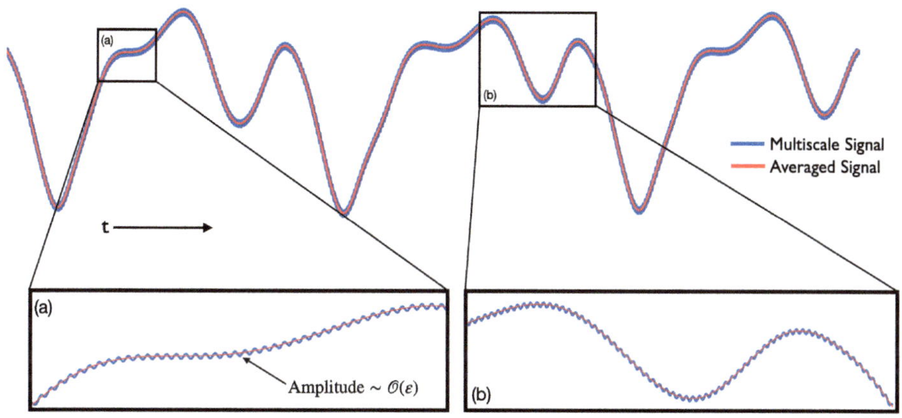

Figure 3.6. *A signal evolving on multiple distinct timescales. Over short periods of time the signal appears periodic, as shown in the zoom windows* (a) *and* (b)*, while over longer timescales a slow drift emerges. Typically the amplitude of oscillation in the fast dynamics scales with the scale separation parameter* $\varepsilon > 0$*. When* ε *is small, the theory of averaging in dynamical systems seeks to remove the fast timescale oscillations, leaving only the slow drifting motion (red).*

The previous example illustrates the basics of averaging, with many powerful extensions going far beyond the scope of this book, such as allowing for (3.76) being almost periodic in t instead [113]. From a computational standpoint, averaging methods offer an advantage to numerical simulation since simulating the full system (3.76) requires small step sizes to resolve the fast-scale oscillations. However, small step sizes require a lot of computations and time to see the slow-scale drift emerge. With the averaged system (3.77) one can easily resolve the slow-scale drifting behavior with larger steps that jump over the predictable fast-scale oscillations.

What is important to note with the theory of averaging is that implementing it requires governing equations. If we are only given a signal but no governing equations, the question becomes, *How can we learn the slow timescale drift independent of the fast oscillations*? That is, suppose we are given a signal $\Phi(t, \varepsilon t)$ that is $T > 0$ periodic in the first component. Applications of model discovery algorithms, such as SINDy, should not be expected to perform well in discovering the governing equations of the signal outright due to the timescale disparity, as illustrated in [40, section III]. Here we use SINDy to define a flow map that iterates through periods of the fast timescale, resulting in a form of *data-driven averaging*.

As the signal is periodic in the fast dynamics, we track it after each period by defining

$$x_n := \Phi(0, \varepsilon nT), \quad n = 0, 1, 2, 3, \ldots, \tag{3.81}$$

where we used $\Phi(nT, \varepsilon nT) = \Phi(0, \varepsilon nT)$ for all $n \geq 0$ by periodicity. The strategy is to use SINDy to discover a flow map which governs

$$x_{n+1} = F(x_n). \tag{3.82}$$

The function F describes the discretized slow-scale drift, which, when $\varepsilon > 0$ is small enough, can be used to recover the continuous-time system that governs the drift. Indeed, supposing that the slow drift of the averaged system is governed by the differential equation

$$\partial_2\Phi(0, t) = G(\Phi(0, t)), \tag{3.83}$$

where ∂_2 represents the partial derivative with respect to the second (slow) component, a Taylor

Table 3.2. *Discovered flow maps* (3.82) *and averaged dynamics* (3.86) *from data generated by* (3.78) *for varying* ε.

ε	Flow map $F(x)$	Averaged dynamics $G(x)$
0.0001	$\texttt{1.00009}x - \texttt{0.00009}x^2$	$\texttt{0.99967}x - \texttt{0.99977}x^2$
0.0005	$\texttt{1.00050}x - \texttt{0.00049}x^2$	$\texttt{0.99754}x - \texttt{0.99838}x^2$
0.001	$\texttt{1.00100}x - \texttt{0.00099}x^2$	$\texttt{0.99605}x - \texttt{0.99709}x^2$
0.005	$\texttt{1.00500}x - \texttt{0.00499}x^2$	$\texttt{0.99873}x - \texttt{0.99953}x^2$
0.01	$\texttt{1.00970}x - \texttt{0.00976}x^2$	$\texttt{0.97378}x - \texttt{0.97625}x^2$
0.05	$\texttt{1.04730}x - \texttt{0.04768}x^2$	$\texttt{0.94602}x - \texttt{0.95365}x^2$
0.1	$\texttt{1.09130}x - \texttt{0.09272}x^2$	$\texttt{0.91254}x - \texttt{0.92724}x^2$

expansion of $x_{n+1} = \Phi(0, \varepsilon(n+1)T)$ about $t = nT$ gives

$$x_{n+1} = \Phi(0, \varepsilon nT) + \varepsilon T \partial_2 \Phi(0, nT) + \mathcal{O}(\varepsilon^2) = x_n + \varepsilon T G(x_n) + \mathcal{O}(\varepsilon^2). \tag{3.84}$$

Furthermore, using the discovered mapping (3.82) gives

$$F(x) = x + \varepsilon T G(x) + \mathcal{O}(\varepsilon^2), \tag{3.85}$$

so, up to order ε^2 corrections, we can identify the averaged dynamics G from F by

$$G(x) = \frac{F(x) - x}{\varepsilon T}. \tag{3.86}$$

Hence, we see that when ε is small, we can recover the continuous-time slow drifting in G from the flow map (3.82). The expansion (3.85) also gives some insight into the choice of the sparsity parameter in the application of SINDy. Since we are only interested in the leading-order terms in ε, we should expect to use a sparsity parameter $\lambda \leq \varepsilon T$. This will help to ensure that the $\mathcal{O}(\varepsilon)$ terms are discovered in the flow map (3.82) while removing unnecessary lower-order ones.

Example 3.9. Let us apply the method of data-driven averaging to a signal governed by (3.79) from the previous example. Results are reported in Table 3.2 for increasing values of the scale separation parameter ε. Both the discovered flow map $F(x)$ and the associated averaged dynamics $G(x)$, per (3.86), are reported. All cases take the sparsity parameter to be $\lambda = \varepsilon^2$, and data is gathered from a single trajectory with a random initial condition in the interval $[0, 4]$.

Notice that as ε increases, the discovered averaged dynamics $G(x)$ begin to differ significantly from the true slow dynamics $x - x^2$ in (3.79). Further, all discovered models differ from the true slow dynamics up to $\mathcal{O}(\varepsilon)$ corrections in the coefficients. These deviations come from the choice of the sparsity parameter since the flow map $F(x)$ can only be correct up to $\mathcal{O}(\varepsilon^2)$ corrections, which we divide off a factor of ε to get $G(x)$. Thus, the method should be expected to perform best in cases when the scale separation is very large, i.e., ε is very small.

Averaging data is applicable to signals beyond those that appear to come from systems of the form (3.76). That is, the typical application of averaging is to a signal whose fast dynamics are micro-oscillations; i.e., their amplitude is proportional to $\varepsilon \ll 1$. Since the data-driven averaging method presented in this section does not require governing equations, it can be applied to any signal $\Phi(t, \varepsilon t)$, even if the amplitude of the fast component is large compared to ε.

It is important to recognize the form of signals that data-driven averaging can be applied to. One such example is a signal that is approximately a linear combination of periodic signals with significantly different periods of oscillation, as we explored previously in Example 2.2. For

Algorithm 3.3 Data-Driven Averaging

1. Sample a signal $x(t)$ and perform sliding window DMD (Algorithm 2.2) to identify dominant fast-scale period $T > 0$
2. Coarse-grain signal by tracking after each fast period $x_n = x(nT)$
3. Apply SINDy (Algorithm 3.1) to identify flow map $x_{n+1} = F(x_n)$
4. Approximate averaged ODE by $G(x) = [F(x) - x]/(\varepsilon T)$

example, we consider a signal of the form

$$x(t) = P_1(t) + P_2(t), \tag{3.87}$$

where P_1 and P_2 are periodic with periods T_1 and T_2, respectively. If $T_2 \gg T_1$, we have that over one period of P_2 numerous periods of P_1 have elapsed. This naturally leads to the scale separation parameter

$$\varepsilon := T_1/T_2 \ll 1. \tag{3.88}$$

Thus, coarse-graining the signal by tracking at each multiple of T_1 allows for the application of data-driven averaging to discover the governing equations for P_2 via (3.86) and the value of ε from (3.88). Importantly, applications to signals of the form (3.87) do not require that the amplitude of oscillation in P_1 be small. Finally, we remind the reader that using the sliding window DMD method of Section 2.2, we can identify T_1 and T_2 directly from data, thus posing no problem for applications of data-driven averaging to signals of the form (3.87). Algorithm 3.3 summarizes the full data-driven averaging method in conjunction with sliding window DMD.

Example 3.10. Periodicity on multiple timescales exemplified in (3.87) has been observed in the motion of the planets Saturn and Jupiter as they orbit about the sun. For this example we will consider time-series data generated from a three-body problem in the heliocentric frame of reference that accounts for pairwise gravitational attraction between only the three most massive objects in our solar system: the sun, Jupiter, and Saturn. Data is generated with a symplectic integrator [91, 271] over 10^6 years with realistic initial conditions and masses for each body.

Restricting the motion of each planet to its orbital plane about the sun results in a signal approximately taking the form (3.87). For both Jupiter and Saturn, P_1 describes the primary elliptic motion of each planet about the sun, while P_2 captures the relatively weaker gravitational interaction between the planets which induces a wobble observed on long timescales. In celestial dynamics this wobbling motion is referred to as the *eccentricity* of the orbit. The sliding window DMD technique of Section 2.2 finds that $T_1 = 11.86$ years for Jupiter and $T_1 = 29.5$ years for Saturn for the test data, providing a close approximation of astronomical findings on their orbital periods. For both Saturn and Jupiter sliding window DMD finds that $T_2 = 46800$ years [77].

Our signal will be the motion of Jupiter in its orbital plane, with coordinates denoted $(x_1^J, x_2^J) \in \mathbb{R}^2$. We gather data at every $T_1 = 11.86$ years, while from (3.88) we have $\varepsilon = 2.6 \times 10^{-4}$. Since $T_1\varepsilon > 10^{-3}$, we will follow the discussion above and take the sparsity parameter to be $\lambda = 10^{-3}$. Smaller values of λ have the potential to introduce spurious terms into the model, as we are already dealing with very small values. The resulting SINDy flow map (3.82) is discovered as

$$\begin{bmatrix} x_1^J \\ x_2^J \end{bmatrix}_{n+1} = \begin{bmatrix} \texttt{0.0036} \\ -\texttt{0.0018} \end{bmatrix} + \begin{bmatrix} \texttt{0.9999} & \texttt{0.0114} \\ -\texttt{0.0114} & \texttt{0.9999} \end{bmatrix} \begin{bmatrix} x_1^J \\ x_2^J \end{bmatrix}_n . \tag{3.89}$$

Thus, using (3.86) we have that the eccentricity dynamics of Jupiter are governed by

$$G(x_1, x_2) = \begin{bmatrix} \mathtt{0.5847} + \mathtt{3.7916}x_2 \\ -\mathtt{1.2063} - \mathtt{3.7983}x_1 \end{bmatrix}. \tag{3.90}$$

Importantly, the eccentricity dynamics have a conserved quantity (see Section 3.5) in that all solutions lie along the implicit curves

$$1.2063x_1 + 1.89915x_1^2 + 0.5847x_2 + 1.8958x_2^2 = C \tag{3.91}$$

for some $C \in \mathbb{R}$. Completing the square allows one to rewrite (3.91) as

$$1.89915(x_1 + 0.31759)^2 + 1.8958(x_2 + 0.1542)^2 = C + 0.2366, \tag{3.92}$$

which allows one to better recognize that the solutions are ellipses. Thus, the eccentricity dynamics have been discovered to be periodic as well. Nearly identical results can be achieved for the eccentricity of Saturn [40] and are left to the accompanying code for the interested reader.

We conclude this section by noting that to perform data-driven averaging we require the fast period of oscillation. Identifying this fast period is a task optimally suited for sliding window DMD from Section 2.2. This is exactly what was done in the previous example to obtain the distinct periods of oscillation. Alternatively, in practice the value of the scale separation parameter ε is not known and cannot be identified from data. This is not an issue though since the slow model $G(x)$ will contain a factor of ε, similar to the averaged equations (3.77) that arise in theory.

3.5 ▪ Discovering Conserved Quantities

Reducing the dimensionality of a complex system can be a rewarding step in understanding its dynamics. One way to do this is to identify the existence of a lower-dimensional invariant manifold. Conserved quantities give exactly such a reduction since they confine trajectories to their lower-dimensional level sets. Familiar examples are Hamiltonian ODEs, where the Hamiltonian function itself provides the desired conserved quantity. The more conserved quantities a system has, the better, with the best case being that there are as many as the degrees of freedom in the system. Such systems are called **integrable** [125], and their dynamics are entirely restricted to the one-dimensional curves lying on the intersection of all conserved quantities. However, identifying conserved quantities is a difficult task with no general method for identifying them.

To initiate the theory of this section, let us start with a dynamical system

$$\dot{x} = f(x), \tag{3.93}$$

governed by a vector field $f : \mathbb{R}^d \to \mathbb{R}^d$ with state-space dimension $d \geq 1$. A smooth function $C : \mathbb{R}^d \to \mathbb{R}$ is a **conserved quantity** of (3.93) if for any solution $x(t)$ the value $C(x(t))$ is constant for all t. Since derivatives of constant functions are zero, a conserved quantity must satisfy the identity

$$0 = \frac{\mathrm{d}}{\mathrm{d}t}C(x(t)) = \nabla C(x(t)) \cdot f(x(t)) \tag{3.94}$$

for all solutions $x(t)$. This means that at every point $x \in \mathbb{R}^d$ the vector field f points in a direction that is orthogonal to the gradient of the function C, so $f(x)$ belongs to the tangent space of $C(x)$

Algorithm 3.4 Discovering Conserved Quantities

1. Gather system snapshots X_n and derivative approximations $\dot{X}_n$ at each snapshot
2. Define a dictionary of differentiable functions $\mathcal{D} = \{\theta_1, \theta_2, \dots, \theta_K\}$ and construct matrix $\Gamma(X, \dot{X})$ as in (2.48)
3. Compute SVD $\Gamma(X, \dot{X}) = U\Sigma V^*$
4. Columns of V corresponding to zero singular values give coefficients of conserved quantities: $C(x) = v_1\theta_1(x) + v_2\theta_2(x) + \cdots + v_K\theta_K(x)$

at every point along $x(t)$. If C is nonconstant on open sets, the condition $\nabla C(x) \cdot f(x) = 0$ for all $x \in \mathbb{R}^d$ is necessary and sufficient for C to be a conserved quantity of the ODE (3.93).

Example 3.11. Example 3.10 discovered the model of the eccentricity of Jupiter's orbit to be

$$G(x_1, x_2) = \begin{bmatrix} 0.5847 + 3.7916x_2 \\ -1.2063 - 3.7983x_1 \end{bmatrix}, \tag{3.95}$$

having conserved quantity $C(x_1, x_2) = 1.2063x_1 + 1.89915x_1^2 + 0.5847x_2 + 1.8958x_2^2$. We verify this by checking $\nabla C \cdot G = 0$ for all $(x_1, x_2) \in \mathbb{R}^2$. Indeed, the gradient of C is

$$\nabla C = \begin{bmatrix} 1.2063 + 3.7983x_1 \\ 0.5847 + 3.7916x_2 \end{bmatrix}, \tag{3.96}$$

and so using (3.95) one may check that $\nabla C \cdot G = 0$, so $C(x_1, x_2) = 1.2063x_1 + 1.89915x_1^2 + 0.5847x_2 + 1.8958x_2^2$ is a conserved quantity of (3.11).

One should notice the presence of the generator of the continuous-time Koopman operator (2.81) in (3.94). Conserved quantities are themselves system observables, prescribed by lying in the nullspace of the Koopman generator. Thus, conserved quantities can be identified using the same data-driven techniques that approximate Koopman eigenfunctions from data.

As with EDMD and SINDy previously, we automate the search for conserved quantities by solving a linear algebra problem. The idea put forth in [136] uses EDMD to identify conserved quantities from data expanded in the span of a dictionary. Take

$$\mathcal{D} = \{\theta_1, \theta_2, \dots, \theta_K\}, \qquad K \geq 1, \tag{3.97}$$

so that $\theta_k : \mathbb{R}^d \to \mathbb{R}$, and expand

$$C(x) = \xi_1\theta_1(x) + \xi_2\theta_2(x) + \cdots + \xi_K\theta_K(x) \tag{3.98}$$

for some $\Xi = [\xi_1 \cdots \xi_K]^\intercal \in \mathbb{R}^K$. Use (3.94) to see that if C is a conserved quantity, then

$$\xi_1\nabla\theta_1(x) \cdot f(x) + \xi_2\nabla\theta_2(x) \cdot f(x) + \cdots + \xi_K\nabla\theta_K(x) \cdot f(x) = 0. \tag{3.99}$$

Suppose that we are given data $X \in \mathbb{R}^{d\times N}$ from the state space and that we have estimates of the derivative at each data point, $\dot{X} \in \mathbb{R}^{d\times N}$. These estimates may come from finite-difference approximations, or, if we have access to the governing equations (3.93), we may apply f to each column of X to get to $\dot{X} = f(X)$. The data matrices X and $\dot{X}$ allow for (3.99) to be recast as the linear algebra problem

$$\Gamma(X,\dot{X})\Xi = 0, \tag{3.100}$$

where

$$\Gamma(X,\dot{X}) := \begin{bmatrix} \nabla\theta_1(X_1)\cdot\dot{X}_1 & \nabla\theta_2(X_1)\cdot\dot{X}_1 & \cdots & \nabla\theta_K(X_1)\cdot\dot{X}_1 \\ \nabla\theta_1(X_2)\cdot\dot{X}_2 & \nabla\theta_2(X_2)\cdot\dot{X}_2 & \cdots & \nabla\theta_K(X_2)\cdot\dot{X}_2 \\ \vdots & \vdots & \ddots & \vdots \\ \nabla\theta_1(X_N)\cdot\dot{X}_N & \nabla\theta_2(X_N)\cdot\dot{X}_N & \cdots & \nabla\theta_K(X_N)\cdot\dot{X}_N \end{bmatrix} \in \mathbb{R}^{N\times K}, \tag{3.101}$$

where X_n and $\dot{X}_n$ denote the columns of X and $\dot{X}$, respectively, for all $n = 1,\dots,N$.

System (3.100) frames discovery of conserved quantities from data as a nullspace identification problem for $\Gamma(X,\dot{X})$. In practice we expect that $N \geq K$ since when $N < K$, the rank-nullity theorem guarantees the existence of a nontrivial nullspace, whose elements may not be related to conserved quantities. Furthermore, even in the case that $N \geq K$, it is possible for the nullspace of $\Gamma(X,\dot{X})$ to be two- or more dimensional. In this case we have infinitely many conserved quantities formed from linear combinations of the nullspace vectors. This is a consequence of the linearity of (3.94) with respect to C. We identify linearly independent conserved quantities through identifying an orthogonal basis for the nullspace of $\Gamma(X,\dot{X})$ using the SVD. Indeed, the right singular vectors (columns of V) corresponding to zero singular values provide an orthogonal basis for the nullspace of $\Gamma(X,\dot{X})$, for which we can further introduce sparsity constraints on these basis vectors using sparse SVD [293] or the alternating direction method [231]. The method is summarized in Algorithm 3.4 for clarity.

Example 3.12. To illustrate discovering conserved quantities from data, we follow the example from [136], where the original method was presented. Consider the three-dimensional system

$$\begin{aligned} \dot{x}_1 &= x_2x_3, \\ \dot{x}_2 &= -2x_1x_3, \\ \dot{x}_3 &= x_1x_2, \end{aligned} \tag{3.102}$$

which has two conserved quantities, given by

$$\begin{aligned} L(x_1,x_2,x_3) &= \frac{1}{2}(x_1^2 + x_2^2 + x_3^2), \\ H(x_1,x_2,x_3) &= \frac{1}{2}(x_1^2 + 2x_2^2 + 3x_3^2). \end{aligned} \tag{3.103}$$

Here L is the momentum, which confines the trajectories to the surface of spheres in 3-space, while H is the total kinetic energy (the Hamiltonian), which confines trajectories to the surface of an ellipsoid in 3-space. Together L and H give that momentum and energy of (3.102) are conserved, and so trajectories are uniquely identified by intersections of these spheres and ellipsoids coming from the level surfaces of L and H. Figure 3.7 provides several trajectories on the surface $L(x_1,x_2,x_3) = \frac{1}{2}$ and $H(x_1,x_2,x_3) = \frac{1}{2}$ for visualization.

Synthetic data is generated from 20 randomly drawn initial conditions in the cube $[-5,5]^3$, each of which is integrated from $t = 0$ to $t = 5$ with a step size $h = 0.01$. The method is implemented with a dictionary of monomials in x_1, x_2, x_3 up to degree 3, while $\dot{X}$ is estimated using the right-hand side of (3.102), i.e., $\dot{X} = f(X)$. The resulting data matrix $\Gamma(X,\dot{X})$ has exactly two zero singular values, while the remaining singular values are at least $\mathcal{O}(10^2)$, leaving no ambiguity to the fact that the system has exactly two conserved quantities. The right singular vectors

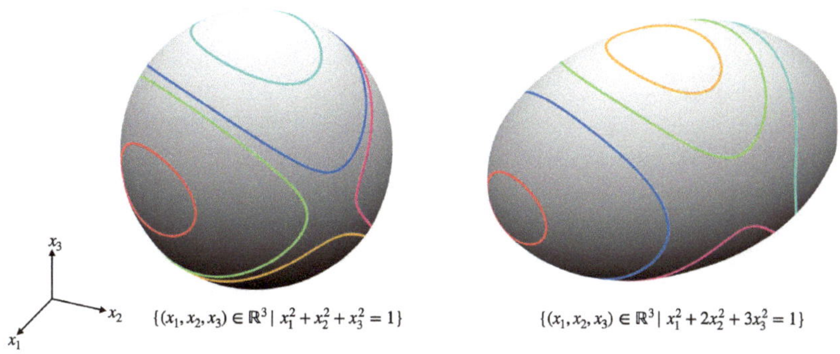

Figure 3.7. *Several trajectories of* (3.102) *lying on the surfaces (left)* $L(x_1, x_2, x_3) = \frac{1}{2}$ *and (right)* $H(x_1, x_2, x_3) = \frac{1}{2}$, *defined in* (3.103). *Since there are two distinct conserved quantities, solutions are uniquely identified as lying on their zero- or one-dimensional intersections.*

of $\Gamma(X, \dot{X})$ associated to these zero singular values provide a basis for the two-dimensional space of conserved quantities of (3.102) as

$$\begin{aligned} C_1(x_1, x_2, x_3) &= \texttt{0.6013}x_1^2 + \texttt{0.5770}x_2^2 + \texttt{0.5527}x_3^2, \\ C_2(x_1, x_2, x_3) &= -\texttt{0.6868}x_1^2 + \texttt{0.0199}x_2^2 + \texttt{0.7266}x_3^2. \end{aligned} \tag{3.104}$$

One can readily check that

$$\begin{aligned} L(x_1, x_2, x_3) &= \texttt{0.0298}C_1(x_1, x_2, x_3) + \texttt{0.8655}C_2(x_1, x_2, x_3), \\ H(x_1, x_2, x_3) &= \texttt{0.7663}C_1(x_1, x_2, x_3) + \texttt{1.7067}C_2(x_1, x_2, x_3), \end{aligned} \tag{3.105}$$

thus giving that the "physical" conserved quantities L and H lie in the span of the conserved quantities discovered from data.

We conclude this section by remarking on the discrete-time characterization of discovering conserved quantities. The presentation in this section was focused on continuous time, as it is in continuum models from physics where one typically encounters conserved quantities for the first time. Nonetheless, a conserved quantity of a discrete-time process $x_{n+1} = F(x_n)$ is a function $C : \mathbb{R}^d \to \mathbb{R}$ that satisfies $C(F(x)) = C(x)$ for all $x \in \mathbb{R}^d$. Indeed, this shows that the output of C does not change along orbits generated of F.

Again we recognize the action of the discrete-time Koopman operator $\mathcal{K}C = C \circ F$ on the observable C. Thus, conserved quantities in discrete time are Koopman eigenfunctions with eigenvalue 1 of the Koopman operator. Equivalently, conserved quantities in discrete time lie in the nullspace of $\mathcal{K} - I$, where I is the identity operator. Thus, the above procedure for identifying conserved quantities from data is readily transferred to discrete time in much the same way as SINDy or any other dictionary-based technique.

3.6 ▪ Discussion and Extensions

Model discovery from data is one of the most important ongoing applications of data-driven methods. It holds great promise for producing models in the absence of governing laws to derive equations of motion from. Although promoting sparsity can help to mitigate the corruption of the data from noisy measurements, there are other robust methods that are designed to handle both noisy and incomplete data [237, 236]. One can also improve model discovery from incomplete measurements of the system by incorporating so-called side information, meaning dynamical features such as boundaries in phase space, Lyapunov functions, or locations of equilibria [6].

Much of this chapter was centered around the SINDy method. This is for the reason that it is a simple and efficient method for performing model discovery with performance guarantees and is versatile enough to be incorporated into a number of different frameworks. For example, SINDy has been applied to discover PDEs [240], boundary value problems [253], chemical reaction models [126], stochastic systems [32], equations with rational functions on the right-hand side [135], and bounded quadratic models [140]. Furthermore, SINDy can be implemented with the Python package PySINDy [70, 141]. However, what remains an open issue for all of these variations of SINDy is the choice of functions that constitute one's dictionary. In Chapter 6 we will return to this issue and demonstrate how neural networks can be employed to discover optimal coordinate transformations to fit data to a limited dictionary.

Chapter 4

Data-Driven Polynomial Optimization

In this chapter we demonstrate the integration of polynomial optimization into data-driven methods to extract system-level information without model identification. This is primarily done through the identification of auxiliary functions. We will see that auxiliary functions constitute a special class of observables that can be used to produce information or statistics about a system. We leverage the ability of EDMD to approximate the action of the Koopman operator on observable functions to produce problems in polynomial optimization that provide insight into a system's dynamics. This chapter provides an exposition of the relationship between the Koopman operator and auxiliary functions to demonstrate how one can extract important information about a complex system without having to build a model for it.

The approach in this chapter allows one to hide the complexity of an underlying system generating given data inside an EDMD matrix. We will see that auxiliary functions allow one to ask questions about the data and receive interpretable results, even if the underlying model is quite complicated or inaccessible. This is contrary to model discovery techniques such as SINDy, which sacrifice accuracy for interpretability. The sacrifice of accuracy means that first using SINDy to discover a model and then analyzing it could lead to drastically inaccurate interpretations of the original data. Furthermore, many model discovery techniques are not applicable to data generated by a random process. Here we will see that the auxiliary function approach is agnostic to the data being produced deterministically or stochastically, meaning that one can apply it in either case to return system-level information without changing the initial setup.

4.1 ▪ Auxiliary Functions

Many important statements about dynamical systems can be proved with scalar-valued functions of the state space whose temporal evolution along trajectories obeys certain pointwise inequalities that imply the desired result. These functions are referred to as **auxiliary functions** and, as we will demonstrate throughout this chapter, can be thought of as a special subset of the observables we have encountered with Koopman operator theory. A frequently encountered example of an auxiliary function is a Lyapunov function, which we begin this section by reviewing.

Example 4.1. For a smooth $f : \mathbb{R}^d \to \mathbb{R}^d$ satisfying $f(0) = 0$, we consider the ODE

$$\dot{x} = f(x). \tag{4.1}$$

A smooth function $V : \mathbb{R}^d \to \mathbb{R}$ is called a **Lyapunov function** [181] if it satisfies

1. $V(0) = 0$ and $V(x) \geq 0$ for all $x \in \mathbb{R}^d \setminus \{0\}$ (positive semidefinite);

2. $\nabla V(x) \cdot f(x) \leq 0$ for all $x \in \mathbb{R}^d$ (nonincreasing along trajectories);

3. $V(x) \to \infty$ as $\|x\| \to \infty$ (radially unbounded).

Conditions (1) and (2) imply that solutions $x(t)$ of (4.1) satisfy $0 \leq V(x(t)) \leq V(x(0))$ for all $t \geq 0$, meaning trajectories cannot leave the closed set $\{x \in \mathbb{R}^d : V(x) \leq V(x(0))\}$, which is compact by condition (3). Finding a Lyapunov function guarantees **Lyapunov stability** of the equilibrium $x = 0$, meaning that trajectories starting close to $x = 0$ must remain close for all $t \geq 0$. Moreover, LaSalle's invariance principle [163] provides that all trajectories of (4.1) will limit as $t \to \infty$ to the set of complete trajectories of (4.1) within the set $\{x \in \mathbb{R}^d : \nabla V(x) \cdot f(x) = 0\}$.

Conditions (1) and (2) can be strengthened to guarantee that every trajectory $x(t)$ to (4.1) converges to the equilibrium $x = 0$ as $t \to \infty$. That is, $x = 0$ is **globally asymptotically stable** if there exists a Lyapunov function $V : \mathbb{R}^d \to \mathbb{R}$ satisfying

1*. $V(0) = 0$ and $V(x) > 0$ for all $x \in \mathbb{R}^d \setminus \{0\}$ (positive definite);

2*. $\nabla V(x) \cdot f(x) < 0$ for all $x \in \mathbb{R}^d \setminus \{0\}$ (decreasing along trajectories);

3. $V(x) \to \infty$ as $\|x\| \to \infty$ (radially unbounded).

This can be seen as a direct consequence of LaSalle's invariance principle, as the set $\{x \in \mathbb{R}^d : \nabla V(x) \cdot f(x) = 0\}$ now contains only $x = 0$.

To illustrate global asymptotic stability, we consider the planar ODE

$$\begin{aligned} \dot{x}_1 &= -x_2 - x_1^3, \\ \dot{x}_2 &= x_1^5. \end{aligned} \tag{4.2}$$

This system has a Lyapunov function given by $V(x_1, x_2) = x_1^6 + 3x_2^2$. Indeed, conditions (1) and (3) are readily verified, while checking condition (2) leads to

$$\nabla V(x_1, x_2) \cdot f(x_1, x_2) = (6x_1^5, 6x_2) \cdot (-x_2 - x_1^3, x_1^5) = -6x_1^8, \tag{4.3}$$

which demonstrates that $V(x_1, x_2) \leq 0$ for all $(x_1, x_2) \in \mathbb{R}^2$. Further, note that while (1*) is satisfied, (2*) is not, as $V(x_1, x_2) \cdot f(x_1, x_2) = 0$ anytime $x_1 = 0$, regardless of the value of x_2. Thus, LaSalle's invariance principle guarantees that every trajectory of (4.2) limits to an element of the set of trajectories that remain on the line $x_1 = 0$ for all t.

One possible limiting trajectory is the equilibrium $(x_1, x_2) = (0, 0)$, but are there any others? Notice that if $x_1(t_0) = 0$ for some t_0, then from the ODE (4.2) we have $\dot{x}_1(t_0) = -x_2(t_0)$. If $x_2(t_0) = 0$, we are at the equilibrium point and have $x_1(t) = x_2(t) = 0$ for all t, while if $x_2(t_0) \neq 0$, then $\dot{x}_1(t_0) \neq 0$, giving that $x_1(t) \neq 0$ for some interval of t values slightly to the right of t_0. Therefore, the only trajectory of (4.2) that can satisfy $\nabla V(x_1, x_2) \cdot f(x_1, x_2) = 0$ for all t is the equilibrium $(0, 0)$. So, LaSalle's invariance principle gives that all trajectories of (4.2) must approach the origin as $t \to \infty$. This demonstrates that conditions (1^*) and (2^*) above are sufficient but not necessary to conclude global asymptotic stability of an equilibrium.

Of the three conditions that determine a Lyapunov function, it is condition (2) that deserves the most interrogation. The term $\nabla V \cdot f$ is the critical component for determining auxiliary functions associated to ODEs, not just Lyapunov functions. To see where it comes from, suppose that $x(t)$ is a solution to the differential equation (4.1). Then, much like the Koopman operator framework introduced in Section 2.4, we are *observing* the evolution of the trajectory inside a candidate auxiliary function, $V(x(t))$. Differentiating with respect to time gives

$$\frac{\mathrm{d}}{\mathrm{d}t} V(x(t)) = \nabla V(x(t)) \cdot \dot{x}(t) = \nabla V(x(t)) \cdot f(x(t)). \tag{4.4}$$

Thus, evaluating $\nabla V \cdot f$ at $x(t)$ provides a formula for the time derivative of the observation $V(x(t))$. With Lyapunov functions, condition (2) guarantees that the auxiliary function V is nonincreasing along trajectories of the system, meaning that we observe the value of $V(x(t))$ to never go up as time goes on.

We saw previously that $\nabla V \cdot f$ is the **Lie derivative** associated to an autonomous ODE. In general, Lie derivatives describe the change in an observable along the direction of the flow of a system. Precisely, they are directional derivatives associated to a vector field. We review some examples of Lie derivatives (denoted $\mathcal{L}$ throughout) for important classes of dynamical systems:

- (*Ordinary Differential Equations*) Given a well-posed ODE $\dot{x} = f(t,x)$ and a differentiable observable function $\phi(t,x)$, the Lie derivative is given by

$$\mathcal{L}\phi(t,x) = \partial_t\phi(t,x) + \nabla_x\phi(t,x) \cdot f(t,x). \tag{4.5}$$

- (*Discrete-Time Dynamical Systems*) For a discrete-time iterative process $x_{n+1} = f(n, x_n)$, the Lie derivative acts on observables $\phi(n,x)$ by

$$\mathcal{L}\phi(n,x) = \phi(n+1, f(n,x)) - \phi(n,x). \tag{4.6}$$

- (*Stochastic Maps*) Suppose that a discrete-time stochastic process $x_{n+1} = f(\omega(n), n, x_n)$ is such that the function $\omega \mapsto f(\omega,\cdot,\cdot)$ takes a random variable from a probability space $(\Omega, \mathcal{F}, \pi)$ to a map from $\mathbb{N} \times \mathbb{R}^d$ to $\mathbb{R}^d$. The Lie derivative acts on observables $\phi(n,x)$ by

$$\mathcal{L}\phi(n,x) = \int_\Omega \phi(n+1, f(\omega, n, x))\mathrm{d}\pi(\omega) - \phi(n,x). \tag{4.7}$$

- (*Stochastic Differential Equations*) Consider the stochastic differential equation

$$\mathrm{d}x = f(t,x)\mathrm{d}t + g(t,x)\mathrm{d}W(t), \tag{4.8}$$

  where $f : \mathbb{R}_+ \times \mathbb{R}^d \to \mathbb{R}^d$, $g : \mathbb{R}_+ \times \mathbb{R}^d \to \mathbb{R}^{d\times k}$, and $\mathrm{d}W(t)$ is a k-dimensional Wiener process. For a twice-differentiable observable function $\phi(t,x)$ we have the Lie derivative

$$\mathcal{L}\phi(t,x) = \partial_t\phi(t,x) + \nabla_x\phi(t,x) \cdot f(t,x) + \frac{1}{2}\langle g(t,x)g(t,x)^\intercal, \nabla_x^2\phi(t,x)\rangle, \tag{4.9}$$

  where $\nabla_x^2\phi(t,x)$ is the $d\times d$ dimensional Hessian matrix with respect to x of the observable ϕ and the angled brackets indicate the Frobenius inner product, which treats matrices as vectors and then applies the standard inner product.

In the list above we have only enumerated the Lie derivatives for some of the standard examples in dynamical systems. In general, the Lie derivative of a discrete-time system is

$$\mathcal{L}\phi(n,x) := \mathbb{E}[\phi(n+1, x_{n+1})|\, x_n = x] - \phi(n,x) \tag{4.10}$$

and in continuous time is

$$\mathcal{L}\phi(t,x) := \lim_{h\to 0^+} \frac{\mathbb{E}[\phi(t+h, x(t+h))|\, x(t) = x] - \phi(t,x)}{h}, \tag{4.11}$$

where $\mathbb{E}$ denotes an expectation. Of course, if the system is deterministic, the expectation drops out, and we simply have pointwise evaluation, as can be observed in the examples above. Importantly, conditions on auxiliary functions that involve Lie derivatives in one context can be exchanged for the more general definitions in discrete or continuous time to apply to a broader range of systems. For example, Lyapunov functions can be identified for any of the four examples enumerated above by simply replacing condition (2) with $\mathcal{L}V \leq 0$ and the appropriate Lie derivative $\mathcal{L}V$ for the system dynamics. In all cases, the Lie derivative captures the fact that the output of the Lyapunov function along trajectories is expected to decrease as time goes on. Conditions (1) and (3) would remain the same, as they are not specific to the flow of the system.

Now that we have a better understanding of how system dynamics are encoded through the Lie derivative of observables, we consider another example for the application of auxiliary functions.

Example 4.2. A problem at the heart of ergodic theory is determining long-time averages of continuous observables [187, 226]. Here we demonstrate how auxiliary functions can aid in this this task. We start with a trajectory $x(t)$ to a dynamical system with initial condition $x(0) = x_0$. Given a quantity of interest $\Phi : \mathbb{R}^d \to \mathbb{R}$, the long-time average of this quantity along solutions to the dynamical system is given by

$$\overline{\Phi}(x_0) = \limsup_{T\to\infty} \frac{1}{T}\int_0^T \Phi(x(t))\mathrm{d}t. \tag{4.12}$$

While our exposition is demonstrated for deterministic, continuous-time processes, this auxiliary function framework can be applied equally to stochastic or discrete-time processes as well. For example, if $x(t)$ is given at discrete time points, then the integral in (4.12) becomes a sum. Similarly, if $x(t)$ is generated by a stochastic process, time averages are replaced by stationary expectations [158, 159]. Nonetheless, all of these situations lead to the same formula at the end of this example, which simply requires the appropriate Lie derivative to be used.

By definition, the averaged quantity $\overline{\Phi}(x_0)$ is dependent on the initial condition used to initiate the trajectory $x(t)$. Then, if we are given a set for which initial conditions are drawn from B, we may ask what the largest possible time average is when starting an initial condition in B:

$$\overline{\Phi}^* := \sup_{x_0\in B} \overline{\Phi}(x_0). \tag{4.13}$$

If B is compact and positively invariant for the flow of the dynamical system, then not only is $\overline{\Phi}^*$ finite, but it is attained by (at least) one initial condition $x_0 \in B$ [273]. Under such conditions we can guarantee existence, but the question remains as to what $\overline{\Phi}^*$ is.

We can bound $\overline{\Phi}^*$ using auxiliary functions. Indeed, given a smooth auxiliary function $V : B \to \mathbb{R}$, if there exists a $U \in \mathbb{R}$ so that

$$U - \Phi(x) - \mathcal{L}V(x) \geq 0 \qquad \forall x \in B, \tag{4.14}$$

then we have that $\overline{\Phi}^* \leq U$. To see why, take the time average of (4.14) to find that for any $x_0 \in B$ we have

$$\begin{aligned} 0 &\leq \overline{U - \Phi(x_0) - \mathcal{L}V(x_0)} \\ &= U - \overline{\Phi}(x_0) - \overline{\mathcal{L}V}(x_0) \\ &= U - \overline{\Phi}(x_0) - \limsup_{T\to\infty} \frac{V(x(T)) - V(x_0)}{T} \\ &= U - \overline{\Phi}(x_0), \end{aligned} \tag{4.15}$$

where the time average of the Lie derivative is simplified using the fact that it is a directional derivative along the flow. Furthermore,

$$\limsup_{T\to\infty} \frac{V(x(T)) - V(x_0)}{T} = 0 \tag{4.16}$$

since we have assumed that B is compact and forward invariant, meaning that $x(T) \in B$ for all $T \geq 0$, and since V is continuous, it follows that $V(x(T))$ is uniformly bounded over B.

For each choice of auxiliary function V we arrive at a candidate for an upper bound U on $\overline{\Phi}^*$. In fact, [273] makes this more precise by proving that

$$\begin{aligned}\overline{\Phi}^* &= \inf_{V \in C^1(B,\mathbb{R})} \max_{x \in B} [\Phi(x) + \mathcal{L}V(x)] \\ &= \inf_{V \in C^1(B,\mathbb{R})} \{U \in \mathbb{R} : U - \Phi(x) - \mathcal{L}V(x) \geq 0 \; \forall x \in B\}.\end{aligned} \tag{4.17}$$

Thus, optimizing over all smooth auxiliary functions not only provides an upper bound on $\overline{\Phi}^*$ but also returns the least upper bound (the maximum value of $\overline{\Phi}^*$). Using the same principles we can bound $\overline{\Phi}^*$ from below using auxiliary functions V by finding $L \in \mathbb{R}$ so that

$$\Phi(x) + \mathcal{L}V(x) - L \geq 0. \tag{4.18}$$

As with the upper bounds, optimizing over all smooth V gives the greatest lower bound on $\overline{\Phi}^*$.

To see this in action, let us return to the Lorenz system (3.8) of Example 3.1. As is shown in [149, 101], taking the quantity of interest to be $\Phi(x) = x_3^2$, one achieves the upper bound $\overline{x_3^2} \leq (\rho - 1)^2$ using the auxiliary function

$$V(x_1, x_2, x_3) = \frac{1}{\beta}\left[2x_3 - 2\rho x_3 + \frac{1}{\sigma}x_1^2 + x_2^2 + x_3^2\right] \tag{4.19}$$

for all $\beta > 0$. This upper bound is sharp, as it is achieved by either of the steady states $(x_1, x_2, x_3) = (\pm\sqrt{\beta(\rho-1)}, \pm\sqrt{\beta(\rho-1)}, \rho - 1)$. A sharp lower bound is encountered without the aid of an auxiliary function, as $\overline{x_3^2} \geq 0$ is always true, while $\overline{x_3^2} = 0$ is achieved by the origin $(x_1, x_2, x_3) = (0, 0, 0)$. Thus, we see that although we can bound $\overline{\Phi}^*$ from above and below, these values should not be expected to coincide, as different trajectories of the system may lead to these extremal averages.

We have now seen two applications of auxiliary functions to dynamical systems analysis. Beyond these examples, applications of auxiliary functions abound. In the final section of this chapter we will provide a more complete discussion of their widespread application while introducing them by example over the coming sections. Importantly, all auxiliary functions make use of the Lie derivative somewhere in the list of conditions that they must satisfy. To fully emphasize this to the reader, we consider one more application of auxiliary functions.

Example 4.3. Here we only consider the case of a continuous-time system, $\dot{x} = f(x)$. However, at this point the reader should understand that these results are easily transferable to discrete-time or stochastic applications using the appropriate Lie derivative. We again consider a quantity of interest $\Phi : \mathbb{R}^d \to \mathbb{R}$ but now seek to determine the pointwise extremal behavior of Φ subject to the flow of f over some global attractor $\mathcal{A}$.

Precisely, we seek to bound

$$\Phi_{\mathcal{A}}^+ := \max_{x \in \mathcal{A}} \Phi(x). \tag{4.20}$$

The work [102] demonstrates that $\Phi_{\mathcal{A}}^+$ can be bounded using auxiliary functions by

$$\begin{aligned}\Phi_{\mathcal{A}}^+ \leq \inf_{\substack{\lambda > 0 \\ V \in C^1(\mathbb{R}^d)}} \{U :\; & U - V(x) - \lambda \mathcal{L}V(x) \geq 0 \quad \forall x \in \mathbb{R}^d \\ & \text{and } V(x) - \Phi(x) \geq 0 \quad \forall x \in \mathbb{R}^d\}.\end{aligned} \tag{4.21}$$

To see why this is the case, notice that if $x(t)$ is a solution to the dynamical system, $\mathcal{L}V(x(t)) = \nabla V(x(t)) \cdot f(x(t))$, and the first condition in (4.21), if satisfied, gives

$$\frac{\mathrm{d}}{\mathrm{d}t}[V(x(t) - U] \leq -\frac{1}{\lambda}[V(x(t)) - U]. \tag{4.22}$$

An application of Grönwall's inequality gives $V(x(t)) - U \leq \mathrm{e}^{-t/\lambda}[V(x(0)) - U]$. As $t \to \infty$, the upper bound decays, and $x(t)$ approaches the global attractor $\mathcal{A}$, giving that $V(x) \leq U$ for all $x \in \mathcal{A}$. Coupling this finding with the second condition in (4.21) then gives that

$$\Phi(x) \leq V(x) \leq U \qquad \forall x \in \mathcal{A}, \tag{4.23}$$

providing an upper bound on $\Phi^+_{\mathcal{A}}$. Lower bounds on Φ over $\mathcal{A}$ can similarly be identified using this framework by finding upper bounds on $-\Phi$.

4.2 ▪ Sums of Squares and Semidefinite Programming

In the previous section we motivated the use of auxiliary functions for the study of dynamical systems, while in practice it is difficult to find such functions analytically. Indeed, notice that one must guess a candidate auxiliary function and then determine whether it satisfies the appropriate inequalities to provide the desired result. Moreover, even if one is given an auxiliary function, deciding nonnegativity based on the constraints it must satisfy can be difficult, especially when the state space of the system is high dimensional. However, there are situations where this process can be automated using a computer, such as when the dynamical system has exclusively polynomial nonlinearities. To see how this is done, we start with a general presentation of determining polynomial nonnegativity and then come back to the application to dynamical systems.

Let us start with the basic problem of determining whether a polynomial function $f : \mathbb{R}^d \to \mathbb{R}$ is nonnegative everywhere; i.e., is $f(x) \geq 0$ for all $x \in \mathbb{R}^d$? If the degree of the polynomial is odd, then we can immediately answer the question in the negative, but what if the degree is even? It turns out that answering this equation when the degree is ≥ 4 is NP-hard [204, 8], which should at least informally indicate to the reader that this is a highly nontrivial task. However, a simple sufficient condition to guarantee nonnegativity of a polynomial is if it can be written as a **sum of squares** (SOS), meaning there exists polynomials $g_i(x)$ so that

$$f(x) = \sum_i [g_i(x)]^2. \tag{4.24}$$

Much like ruling out polynomials of odd degree, an SOS representation provides an easy check that the polynomial is nonnegative since trivially each $[g_i(x)]^2 \geq 0$. For example, the Lyapunov function $x_1^6 + 3x_2^2$ from Example 4.1 is seen to be an SOS since $x_1^6 + 3x_2^2 = (x_1^3)^2 + (\sqrt{3}x_2)^2$, which verifies that condition (1) for a Lyapunov function is satisfied.

Although SOS representations provide a simple way to check whether a polynomial is nonnegative, this is not always easy to see right away. For example, is the polynomial

$$p(x_1, x_2) = 13x_1^4 - 6x_1^3x_2 - 4x_1^3 + x_1^2x_2^2 + 10x_1^2 + 12x_1x_2^2 + 4x_2^4 \tag{4.25}$$

nonnegative for all $x_1, x_2 \in \mathbb{R}$? The answer is yes, as it admits the SOS representation [7]

$$p(x_1, x_2) = (x_1 - 2x_1^2)^2 + (3x_1 + 2x_2^2)^2 + (x_1x_2 - 3x_1^2)^2. \tag{4.26}$$

To find an SOS representation for a polynomial we use the fact that $f : \mathbb{R}^d \to \mathbb{R}$ admits an SOS decomposition if and only if there exists a positive semidefinite matrix $P \succeq 0$ and a vector of monomials z in x so that

$$f(x) = z^{\mathsf{T}}Pz. \tag{4.27}$$

To see why, we use the fact that a positive semidefinite matrix can be factored as $P = Q^\mathsf{T} Q$, so

$$f(x) = z^\mathsf{T} P z = \|Qz\|^2 = \sum([Qz]_i)^2, \tag{4.28}$$

providing the SOS representation. Returning to (4.25), we have

$$\begin{aligned} p(x_1,x_2) &= 13x_1^4 - 6x_1^3x_2 - 4x_1^3 + x_1^2x_2^2 + 10x_1^2 + 12x_1x_2^2 + 4x_2^4 \\ &= \begin{bmatrix} x_1 \\ x_2 \\ x_1^2 \\ x_1x_2 \\ x_2^2 \end{bmatrix}^\mathsf{T} \begin{bmatrix} 10 & 0 & -2 & 0 & 6 \\ 0 & 0 & 0 & 0 & 0 \\ -2 & 0 & 13 & -3 & 0 \\ 0 & 0 & -3 & 1 & 0 \\ 6 & 0 & 0 & 0 & 4 \end{bmatrix} \begin{bmatrix} x_1 \\ x_2 \\ x_1^2 \\ x_1x_2 \\ x_2^2 \end{bmatrix}, \end{aligned} \tag{4.29}$$

so that

$$Q = \begin{bmatrix} 1 & 0 & -2 & 0 & 0 \\ 3 & 0 & 0 & 0 & 2 \\ 0 & 0 & -3 & 1 & 0 \end{bmatrix}, \tag{4.30}$$

which leads to the SOS representation (4.26). Importantly, if we are given a polynomial f, determining nonnegativity is now equivalent to identifying the matrix P, which amounts to tuning its coefficients to satisfy a system of *linear* equality constraints. Coupling these linear constraints with the restriction that the P must be positive semidefinite constitutes a **semidefinite program** (SDP) which can be solved numerically using free software packages such as MOSEK [15]. The reader is directed to Appendix C for more background on SDPs.

A major question remains: Do all nonnegative polynomials admit an SOS representation? The answer, like many statements in math, is *sometimes*. Hilbert showed that every nonnegative polynomial admits an SOS representation if and only if $d = 1$, the polynomial is degree 2, or $d = 2$ and the polynomial is quartic [124]. An example of a nonnegative polynomial that is not SOS wasn't found until 1967, when Motzkin used the AM–GM inequality to show that the polynomial $x_1^4x_2^2 + x_1^2x_2^4 - 3x_1^2x_2^2x_3^2 + x_3^6$ is globally nonnegative while also proving that it has no SOS representation [202]. Therefore, it is not always the case that a nonnegative polynomial can be written as an SOS, but every nonnegative polynomial can be approximated to arbitrary accuracy by SOS polynomials [169]. Moreover, with the aid of computational software such as MOSEK, we can easily check the sufficient condition for nonnegativity of a polynomial through an SOS representation, which results in a simple check of feasibility of the resulting SDP.

Bringing this back to dynamical systems, this means that if the given system is polynomial, we can check if a polynomial auxiliary function satisfies the necessary inequalities by asking if there exists an SOS representation for the resulting functions required to be nonnegative. This can be taken even further by incorporating a search for the desired auxiliary functions. In particular, we can fix a degree for our polynomial auxiliary function and allow the coefficients to be tunable parameters that are added into the SDP that identifies the entries for the positive semidefinite matrices to give the SOS representations. Such an SDP is an example of **polynomial optimization**, in this case applied to dynamical systems. We illustrate with the following example.

Example 4.4. We consider the Moore–Greitzer system for a simple jet engine model with controller [201], given by

$$\begin{aligned} \dot{x}_1 &= -x_2 - \frac{3}{2}x_1^2 - \frac{1}{2}x_1^3, \\ \dot{x}_2 &= 3x_1 - x_2. \end{aligned} \tag{4.31}$$

For a fixed degree $\alpha \geq 0$, we search for candidate Lyapunov functions of the form

$$V(x_1,x_2) = \sum_{0 \leq i+j \leq \alpha} c_{ij} x_1^i x_2^j, \tag{4.32}$$

where the $c_{i,j} \in \mathbb{R}$ are tunable coefficients of the polynomial auxiliary function. To satisfy conditions (1*), (2*), and (3) in Example 4.1, which demonstrate global asymptotic stability of the unique equilibrium $(x_1, x_2) = (0,0)$ in (4.31), we put the following conditions on V:

1. $V(x_1, x_2) - x_1^2 - x_2^2$ is SOS;
2. $-\mathcal{L}V - x_1^2 - x_2^2$ is SOS.

Here the Lie derivative is given by

$$\mathcal{L}V = \left(- x_2 - \frac{3}{2}x_1^2 - \frac{1}{2}x_1^3 \right) \frac{\partial V}{\partial x_1} + \left(3x_1 - x_2 \right) \frac{\partial V}{\partial x_2}. \tag{4.33}$$

Satisfying the first condition on V would guarantee that

$$V(x_1, x_2) - x_1^2 - x_2^2 \geq 0 \implies V(x_1, x_2) \geq x_1^2 + x_2^2, \tag{4.34}$$

which means $V(x_1, x_2)$ is strictly positive everywhere except the equilibrium point while also providing the coercivity condition (3). The second condition above similarly gives that the Lie derivative is strictly decreasing away from the equilibrium point, thus satisfying condition (2*).

We use the freely available software YALMIP [177] to reformulate polynomial SOS conditions into the related SDP, which are then solved by MOSEK. A Lyapunov function, if it exists, need not be unique, and so we will impose an objective to minimize the ℓ^1-norm of the coefficients of V to improve interpretability. Attempting to identify a quadratic Lyapunov function ($\alpha = 2$) returns an infeasible SDP. Notice that this does not necessarily imply that no quadratic Lyapunov function exists for system (4.31)—only that one that satisfies the given SOS constraints cannot be identified. However, increasing the degree to $\alpha = 4$ identifies the Lyapunov function

$$\begin{aligned} V(x_1, x_2) =& \texttt{15.24798} x_1^2 - \texttt{4.16469} x_1 x_2 + \texttt{4.75991} x_2^2 + \texttt{8.06101} x_1^2 x_2 \\ & - \texttt{1.11749} x_1 x_2^2 + \texttt{1.52750} x_2^3 + \texttt{1.05488} x_1^4 + \texttt{0.22966} x_1^3 x_2 \\ & + \texttt{1.19017} x_1^2 x_2^2 + \texttt{0.18011} x_2^4. \end{aligned} \tag{4.35}$$

Therefore, with V as above we have numerically verified the global asymptotic stability of the origin in (4.31) by identifying a Lyapunov function using computational software.

Now that we have discussed how to obtain SOS representations that verify global nonnegativity, we turn to the case of only enforcing nonnegativity locally. Suppose we have a polynomial $f : \mathbb{R}^d \to \mathbb{R}$ and we wish to determine if

$$f(x) \geq 0 \qquad \forall x \in S \subsetneq \mathbb{R}^d. \tag{4.36}$$

This again becomes a difficult problem to solve, but in some cases it can be strengthened to the feasibility of an SOS decomposition. One such case is when S is a **semialgebraic set**, meaning that it is defined by a finite sequence of polynomial equations and inequalities. Precisely, let us suppose there exist polynomials $s_1(x), \ldots, s_I(x)$ and $r_1(x), \ldots, r_J(x)$ with $I, J \geq 0$ so that

$$S = \{x \in \mathbb{R}^d : \; s_i(x) \geq 0 \text{ for } i = 1, \ldots, I, \; r_j(x) = 0 \text{ for } j = 1, \ldots, J\}. \tag{4.37}$$

A sufficient condition for $f(x) \geq 0$ on S is to have polynomials $\{\sigma_i\}_{i=1}^I$ and $\{\rho_j(x)\}_{j=1}^J$ so that

$$f(x) - \sum_{i=1}^{I} s_i(x)\sigma_i(x) - \sum_{j=1}^{J} r_j(x)\rho_j(x) \text{ is SOS} \tag{4.38}$$

and

$$\sigma_i(x) \text{ is SOS} \quad \forall i = 1, \ldots, I. \tag{4.39}$$

Indeed, notice that the SOS condition guarantees that (4.38) is nonnegative for every $x \in \mathbb{R}^d$ and in particular when $x \in S$. Moreover, for $x \in S$, $r_j(x)\rho_j(x) = 0$ by definition of S, leaving

$$f(x) - \sum_{i=1}^{I} s_i(x)\sigma_i(x) \geq 0 \implies f(x) \geq \sum_{i=1}^{I} s_i(x)\sigma_i(x) \qquad \forall x \in S. \tag{4.40}$$

We have $\sigma_i(x) \geq 0$ for all $x \in \mathbb{R}^d$ since they are SOS, while $s_i(x) \geq 0$ by definition of S, giving

$$f(x) \geq \sum_{i=1}^{I} s_i(x)\sigma_i(x) \geq 0 \qquad \forall x \in S, \tag{4.41}$$

which proves the nonnegativity of f on S.

We can incorporate the SOS relaxations (4.38) and (4.39) into our search for auxiliary functions to localize to specific regions of phase space. Again the function f may include tunable coefficients coming from the search for the desired auxiliary function while we simultaneously tune coefficients of the σ and ρ functions to satisfy the conditions in (4.38) and (4.39). This again gives a polynomial optimization problem, which is translated to an SDP and solved numerically.

Example 4.5. For a parameter $\mu > 0$ and a natural number $m \in \mathbb{N}$, consider the planar ODE

$$\begin{aligned} \dot{x}_1 &= x_2, \\ \dot{x}_2 &= -\mu x_2 - x_1^m(1 - x_1). \end{aligned} \tag{4.42}$$

Such an equation arises from the search for traveling wave solutions to a scalar reaction diffusion equation [27] (see Example 5.5 for the derivation of a similar equation), while the parameter value $\mu > 0$ represents the speed at which the wave is traveling through the spatial medium. The goal in working with (4.42) is to establish the existence or nonexistence of a homoclinic connection between equilibria $(x_1, x_2) = (1, 0)$ and $(x_1, x_2) = (0, 0)$ with $0 \leq x_1 \leq 1$ and $x_2 \leq 0$ [27, 48, 28, 43]. Moreover, it is known that for each $m \in \mathbb{N}$, there exists a unique $\mu_* > 0$ for which such a heteroclinic connection exists if and only if $\mu \geq \mu_*$.

To establish the existence and/or nonexistence of a heteroclinic connection we will build boundaries in phase space, obtained through an auxiliary function. The reader is referred to [43] for full details, while the key ideas are highlighted in Figure 4.1. The equilibrium point $(1, 0)$ has a one-dimensional unstable manifold that enters into the region of interest, $S = \{(x_1, x_2) : 0 \leq x_1 \leq 1,\ x_2 \leq 0\}$, and can only leave by crossing the line $x_1 = 0$. Thus, for a fixed value of $\mu > 0$ we can attempt to find a curve in phase space that forms a barrier for the unstable manifold and either guarantees that we converge into $(0, 0)$ because the region is compact or prevents such a connection and guarantees that the unstable manifold eventually leaves S through $x_1 = 0$.

To identify trapping boundaries that confirm the existence of a heteroclinic orbit, we search for an auxiliary function $V(x_1, x_2)$ satisfying the following:

$$\begin{aligned} \lambda \mathcal{L}V(x_1, x_2) - V(x_1, x_2) \geq 0 \quad & \forall (x_1, x_2) \in S, \\ -V(x_1, 0) - \varepsilon x_1(1 - x_1) \geq 0 \quad & \forall x_1 \in [0, 1], \\ V(0, 0) = 0,& \end{aligned} \tag{4.43}$$

where $\lambda, \varepsilon > 0$ are hyperparameters to be set by the user. In the above $\mathcal{L}V$ is the Lie derivative associated to the planar system (4.42), while the first equation similarly leads to the Grönwall

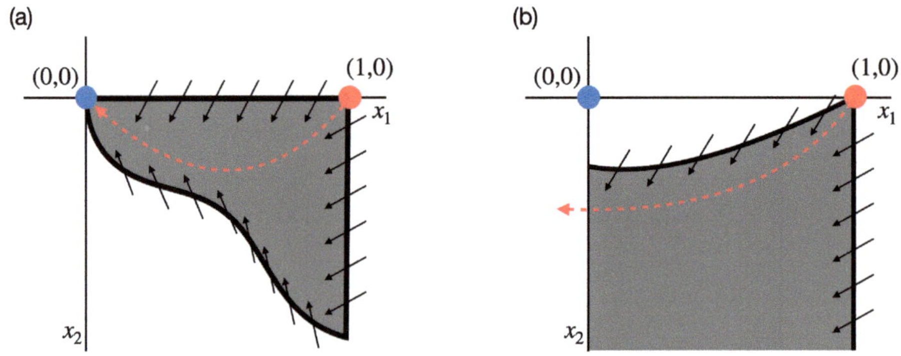

Figure 4.1. *Cartoons of how boundaries are used to prove existence and nonexistence of heteroclinic connections in* (4.42). (a) *An illustration of a trapping region (shaded) guarantees that the unstable manifold (red dashed line) of* $(1,0)$ *connects to* $(0,0)$. (b) *An illustration of a barrier that rules out homoclinic connections because trajectories cannot leave the shaded region without crossing* $x_1 = 0$.

bound (4.22) that implies that the negative-V region is invariant to the dynamics. In particular, the zero level set of V forms the desired trapping boundary, while the second condition in (4.43) guarantees that the $V \leq 0$ region has the geometry illustrated in Figure 4.1(a). The condition $V(0,0) = 0$ puts the equilibrium $(0,0)$ at the pinch point of the trapping region, which in turn guarantees that the unstable manifold of $(1,0)$ is forced toward it.

Using auxiliary functions to construct a barrier in phase space that confirms nonexistence of a heteroclinic orbit is similar. We search for an auxiliary function $V(x_1, x_2)$ satisfying

$$\begin{aligned} \lambda \mathcal{L}V(x_1,x_2) - V(x_1,x_2) \geq 0 &\quad \forall (x_1,x_2) \in S, \\ V(x_1,0) - \varepsilon x_1(1-x_1) \geq 0 &\quad \forall x_1 \in [0,1], \\ V(0,-\varepsilon) = V(1,0) = 0, & \end{aligned} \tag{4.44}$$

again using hyperparameters $\lambda, \varepsilon > 0$. We again have that the negative-V region is invariant in S with $V = 0$ forming an invariant boundary. The second condition in (4.44) enforces that the line $0 \leq x_1 \leq 1$ is not part of the invariant region, leading to the geometry depicted in Figure 4.1(b). The final conditions guarantee that the unstable manifold must enter into the invariant region while also excluding $(0,0)$ from it, meaning that no heteroclinic connection is possible.

Each of the inequalities in (4.43) and (4.44) can be translated into SOS constraints as in (4.38). The set S has no equality constraints, while the inequality constraints are encoded by

$$s_1(x_1,x_2) = x_1(1-x_1) \geq 0, \qquad s_2(x_1,x_2) = -x_2 \geq 0. \tag{4.45}$$

With these SOS constraints, if at a fixed value of $\mu > 0$ we are able to identify an auxiliary function V that satisfies (4.43), then this value of μ provides an upper bound on μ_*. For each computational implementation of the SDPs we can find the smallest μ value at which the existence of a heteroclinic connection can be verified using (4.43). We begin with a large value of μ at which the SOS computation can identify a desired V and a small value of μ at which no such V can be found. We then bisect repeatedly in μ, solving an SOS program at the new value of μ to find the smallest upper bound on μ_* that is achievable. Similarly, for the best lower bound on μ_* we carry out an analogous bisection search in μ to find the largest value for which (4.44) is feasible, after being translated into an SOS program.

Throughout the computations that follow we fix $\varepsilon = 10^{-4}$, as computations show that ε only has an $\mathcal{O}(\varepsilon)$ effect on the upper and lower bounds. The upper and lower bounds have a much

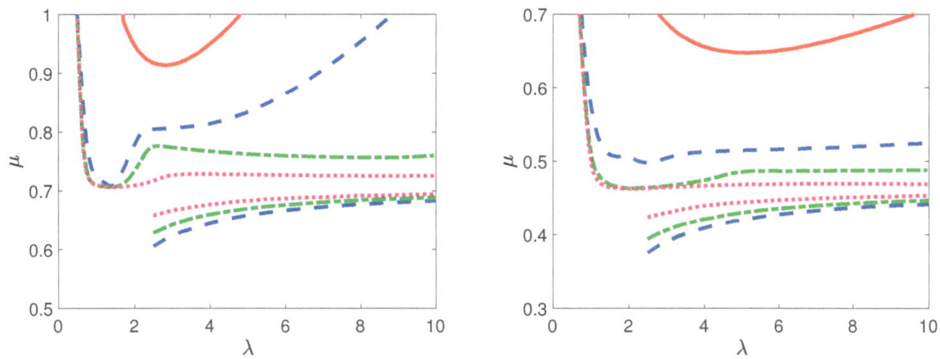

Figure 4.2. *Dependence of upper and lower bounds on* μ_* *in* (4.42) *with (left)* $m = 2$ *and (right)* $m = 3$. *Each line denotes upper and lower bounds computed by satisfying the inequalities* (4.43) *and* (4.44), *respectively, with* $\varepsilon = 10^{-4}$ *and* $\deg(V) = 2$ *(red, solid),* 3 *(blue, dashed),* 4 *(green, dashed-dotted), and* 5 *(pink, dotted). Lower bounds are not shown for* $\lambda < 2.5$ *due to numerical inaccuracy.*

Table 4.1. *Upper and lower bounds on* μ_* *in* (4.42) *computed using* $\deg(V) = 20$ *and* $\varepsilon = 10^{-4}$. *Upper bounds use* $\lambda = 2$, *and lower bounds use* $\lambda = 10^3$.

m	2	3	4	5	6	7	8	9	10
Upper bound	0.708	0.463	0.346	0.278	0.232	0.200	0.176	0.160	0.148
Lower bound	0.707	0.463	0.346	0.277	0.231	0.199	0.174	0.155	0.140

more complex dependence on the other hyperparameter, λ. This dependence is illustrated in Figure 4.2, where we report the best upper and lower bounds on μ_* for $m = 2$ (left) and $m = 3$ (right) for $\lambda \in (0, 10]$ and $\deg(V) = 2, 3, 4, 5$. The degrees of the σ polynomials that are used to localize the inequalities are taken to be $\deg(V) + m$ to align with the degree of $\mathcal{L}V$. There is no known method for anticipating this complex dependence on λ, but we do note two things of importance that seem to be true in general:

1. For a fixed $\lambda > 0$, bounds improve as $\deg(V)$ is increased.
2. As $\deg(V)$ increases, the bounds appear to become insensitive to λ.

Lower bounds appear to be less sensitive to λ and appear to continually improve as λ gets larger; however, it is not clear how one would quantify this analytically.

To conclude this example, we report tight upper and lower bounds over values of m with $\deg(V) = 20$ in Table 4.1. Upper bounds use $\lambda = 2$, and lower bounds use $\lambda = 10^3$. Results are only reported to three decimal places to avoid corrections due to ε. Bounds, especially for larger m, could be improved by sweeping over λ as in Figure 4.2 and/or increasing the degree of V.

4.3 ▪ Auxiliary Functions as Koopman Observables

So far we have been introduced to auxiliary functions and their use in dynamical systems, but the approach to finding them has always required knowing the governing equations. In this section we turn to identifying auxiliary functions directly from data. One may have already noticed the resemblance between the Koopman operator introduced in Chapter 2 and the deterministic Lie derivatives introduced in Section 4.1. The goal in this section is to make this correspondence precise and show that we can leverage EDMD to approximate Lie derivatives acting on dictionaries of observables. The result is a method of extracting information about a system directly from data while circumventing the intermediary step of first performing model discovery.

Let us begin by introducing the Koopman operator more generally than in Chapter 2. Letting $x(t)$ denote the state of a system at time t, the **stochastic Koopman operator** acts on bounded observable functions ϕ by [284, 147, 65]

$$\mathcal{K}_\tau\phi(x) := \mathbb{E}[\phi(x(t+\tau))|\, x(t) = x] \tag{4.46}$$

for some timestep $\tau > 0$. Note that if $x(t)$ is generated by a discrete-time process, then $t, \tau \in \mathbb{N}$. Furthermore, when the dynamics that generate $x(t)$ are deterministic, the expectation in (4.46) drops out, and we are left with pointwise evaluations in ϕ, which in turn coincides with the previously presented Koopman operator from Chapter 2. Importantly, the generator of $\mathcal{K}_\tau$ is the Lie derivative, for which one can now see the presence of the Koopman operator in the definition of the Lie derivative in (4.10) and (4.11). With this correspondence between the Koopman operator and the Lie derivative, auxiliary functions can be thought of as a special class of observable functions within the Koopman operator framework.

In this section we therefore leverage the ability for EDMD to approximate the Koopman operator and then apply methods of semidefinite programming to identify auxiliary functions directly from data. We begin with two dictionaries of polynomial observables, denoted as

$$\mathcal{D}_p = \{p_1(x), p_2(x), \dots, p_\ell(x)\} \tag{4.47}$$

and

$$\mathcal{D}_q = \{q_1(x), q_2(x), \dots, q_m(x)\} \tag{4.48}$$

for positive integers m, ℓ. We will write

$$\vec{p} = (p_1, \dots, p_\ell)^\intercal, \qquad \vec{q} = (q_1, \dots, q_m)^\intercal \tag{4.49}$$

and further assume that $\text{span}\{\vec{p}\} \subseteq \text{span}\{\vec{q}\}$. Thus, there exists a matrix $\Theta \in \mathbb{R}^{\ell\times m}$ so that

$$\vec{p} = \Theta\vec{q}. \tag{4.50}$$

The reason for having two dictionaries will be explained after the presentation of the method. For now, note that these conditions on the library are easily satisfied for typical applications. For example, one would expect to take $\mathcal{D}_p$ to be all monomials in x up to a fixed degree, meaning that $\mathcal{D}_q$ could be all monomials in x up to the same degree or higher. In this case the matrix Θ would simply act to match monomials in $\mathcal{D}_q$ with the same monomial in $\mathcal{D}_p$.

As with our previous data-driven methods, we assume that we have data matrices $X, Y \in \mathbb{R}^{d\times N}$, for which each column of Y represents the same column of X exactly $h > 0$ time units in the future. With these dictionaries and the data, we construct the matrices

$$P = \begin{bmatrix} p_1(Y_1) & p_1(Y_2) & \cdots & p_1(Y_N) \\ p_2(Y_1) & p_2(Y_2) & \cdots & p_2(Y_N) \\ \vdots & \vdots & \ddots & \vdots \\ p_\ell(Y_1) & p_\ell(Y_2) & \cdots & p_\ell(Y_N) \end{bmatrix} \in \mathbb{R}^{\ell\times N} \tag{4.51}$$

and

$$Q = \begin{bmatrix} q_1(X_1) & q_1(X_2) & \cdots & q_1(X_N) \\ q_2(X_1) & q_2(X_2) & \cdots & q_2(X_N) \\ \vdots & \vdots & \ddots & \vdots \\ q_m(X_1) & q_m(X_2) & \cdots & q_m(X_N) \end{bmatrix} \in \mathbb{R}^{m\times N}, \tag{4.52}$$

where X_n and Y_n denote the nth column of matrices X and Y, respectively. The Koopman operator $\mathcal{K}_h$ acting on the span of $\mathcal{D}_p$, as given in (4.46), is then approximated by the linear operator $K_h : \text{span}\{\vec{p}\} \to \text{span}\{\vec{q}\}$ given by

$$K_h := PQ^\dagger \in \mathbb{R}^{\ell\times m}. \tag{4.53}$$

Algorithm 4.1 Approximating Koopman Generators

1. For each snapshot X_n, gather its subsequent snapshot Y_n at $h > 0$ units into the future
2. With dictionaries $\mathcal{D}_p = \{p_1(x), p_2(x), \dots, p_\ell(x)\}$ and $\mathcal{D}_q = \{q_1(x), q_2(x), \dots, q_m(x)\}$ mapped into each other with matrix Θ, construct matrices P as in (4.51) and Q as in (4.52)
3. Approximate the Koopman operator $K_h = PQ^\dagger$
4. Approximate the Lie derivative (the Koopman generator) $L_h = [K_h - \Theta]/h$

Such a formula again comes from identifying a matrix K that minimizes the Frobenius norm $\|P - KQ\|_F$. For every $p = c \cdot \vec{p} \in \text{span}\{\vec{p}\}$ with $c \in \mathbb{R}^\ell$ we set

$$K_h p := c \cdot K_h \vec{q} \in \text{span}\{\vec{q}\}. \tag{4.54}$$

With the above approximation of the Koopman operator, we can approximate the Lie derivative as $L_h : \text{span}\{\vec{p}\} \to \text{span}\{\vec{q}\}$ with

$$L_h := \frac{K_h - \Theta}{h}, \tag{4.55}$$

acting by $L_h p = c \cdot L_h \vec{q}$ for every $p = c \cdot \vec{p} \in \text{span}\{\vec{p}\}$. In discrete time we have $h = 1$, but with continuous-time dynamics we would like to have $h > 0$ small to make L_h a finite-difference approximation of the associated Lie derivative. Work in [42] proves under mild conditions that for any $p = c \cdot \vec{p}$ we have $L_h p \to \mathcal{L}p$, the true Lie derivative, as $N \to \infty$, $h \to 0$ (if the dynamics are continuous), and $m \to \infty$, similar to the convergence of EDMD encountered in Section 2.4. The method is summarized in Algorithm 4.1.

We now have a method of approximating the Lie derivative directly from data, which can thus be used to discover auxiliary functions directly from the data. Prior to performing experiments, we note some aspects of the implementation. First, unless the dynamics of the underlying process are linear, one should expect that $\mathcal{L}p$ for any $p \in \text{span}\{\vec{p}\}$ has higher polynomial degree than p itself. This is exactly what the second dictionary $\mathcal{D}_q$ is attempting to capture, which is of course different from the standard application of EDMD of Section 2.4, where we had $\mathcal{D}_p = \mathcal{D}_q$. Second, much like the presentation in Section 4.1, this data-driven approximation works for both deterministic and stochastic dynamics, requiring no prior adjustments to accommodate how the data is generated. Finally, since $\mathcal{D}_q$ is a dictionary of polynomials, any problem making use of auxiliary functions that can be strengthened to an SOS problem can be approximated using data from the system by replacing the exact Lie derivative with its approximation L_h. Searching for auxiliary functions of increasing polynomial degree to improve results, as was done in Example 4.5, is achieved by increasing the size of the dictionary $\mathcal{D}_p$ in this data-driven framework.

Example 4.6. Consider the planar map

$$\begin{bmatrix} x_1 \\ x_2 \end{bmatrix}_{n+1} = \begin{bmatrix} \frac{3}{10}x_1 \\ -x_1 + \frac{1}{2}x_2 + \frac{7}{18}x_1^2 \end{bmatrix}_n, \tag{4.56}$$

for which the fixed point $(x_1, x_2) = (0, 0)$ is globally asymptotically stable. Our goal is to prove this fact by identifying a Lyapunov function using only data gathered from the system.

Our data matrix $X \in \mathbb{R}^{2\times N}$ is comprised of $N = 10^2$ random (x_1, x_2) values from the box $[-2, 2] \times [-2, 2]$, while $Y \in \mathbb{R}^{2\times N}$ contains the image of each pair under the map (4.56). We will consider $\mathcal{D}_p$ to be all monomials in x_1 and x_2 up to degree α and $\mathcal{D}_q$ to be the monomials up to degree $\beta \geq \alpha$. In practice, we find that taking any $\beta \geq \alpha$ has little effect on the problem, but larger values of α and β may require more data points. The reader is encouraged to experiment with the code that accompanies this example to explore the intricacies of the performance.

As in Example 4.4, we search for a Lyapunov function $V \in \operatorname{span}\{\vec{p}\}$ using the following SOS problem:

1. $V - x_1^2 - x_2^2$ is SOS,
2. $-LV - x_1^2 - x_2^2$ is SOS,

where L is the data-driven approximation of the Lie derivative coming from (4.55). As in Example 4.4, the resulting SDP has the objective to minimize the ℓ^1-norm of the coefficients in V for interpretability. The code is not able to identify a quadratic ($\alpha = 2$) Lyapunov function, no matter the value of β. This is not a feature of the data-driven framework but, as shown in [9, Example 3.1], comes from the fact that no quadratic Lyapunov function for (4.56) exists. Taking instead $\alpha = 4$ and $\beta = 6$ now results in the Lyapunov function

$$\begin{aligned} V(x_1, x_2) = \texttt{3.0815}x_1^2 + \texttt{1.5686}x_1x_2 + \texttt{1.3333}x_2^2 \\ - \texttt{1.3038}x_1^3 + \texttt{0.5428}x_1^2x_2 + \texttt{0.2226}x_1^4, \end{aligned} \tag{4.57}$$

where we have omitted coefficients that are smaller than 10^{-4} for clarity.

Since we have the true dynamics of the system, we can check that the discovered function (4.57) is indeed a Lyapunov function for (4.56). One method of doing so would be taking the Lyapunov function (4.57) and checking that the SOS conditions above are satisfied when we use the true Lie derivative, in this case given by

$$\mathcal{L}V = V\left(\frac{3}{10}x_1, -x_1 + \frac{1}{2}x_2 + \frac{7}{18}x_1^2\right) - V(x_1, x_2). \tag{4.58}$$

One finds that the resulting SDP is infeasible, meaning that an SOS decomposition is not possible using (4.57). However, this does not mean that (4.57) is not a Lyapunov function for (4.56). Indeed, SOS programs only provide *sufficient* conditions for verifying nonnegativity, while also the SOS conditions above are only sufficient conditions to obtain a Lyapunov function.

Alternatively, [42] proposes to use the exact Lie derivative (4.58) and the candidate function (4.57) and then seeks to identify a value of $\varepsilon > 0$ so that the following are true:

1. $V - \varepsilon(x_1^2 + x_2^2)$ is SOS;
2. $-\mathcal{L}V - \varepsilon(x_1^2 + x_2^2)$ is SOS.

Practically, we can formulate the SOS problem as an SDP with the objective to maximize the value of ε for which such SOS decompositions hold. If the maximal value is positive, we indeed have a Lyapunov function, as the conditions above similarly give the desired strict inequalities and coercivity of the function. With the function (4.57) we return $\varepsilon \approx \texttt{0.999994}$, verifying that we have indeed discovered a Lyapunov function from data.

So far we have focused exclusively on deterministic dynamics. However, the general presentation in this chapter allows one to readily apply these methods to stochastic systems as well. Indeed, in this section we were introduced to the stochastic Koopman operator, and in Section 4.1 we were introduced to the related stochastic Lie derivatives. The advantage of the present coupling of EDMD with semidefinite programming is that it applies equally to deterministic and stochastic data, requiring no prior adjustments to accommodate the type of data generation.

Example 4.7. We will consider data generated by the stochastic logistic map, given by

$$x_{n+1} = \lambda_n x_n(1 - x_n), \tag{4.59}$$

where $x_n \in [0, 1]$ and λ_n is drawn from the uniform distribution on $[0, 4]$ at each $n \geq 0$. In this case the Lie derivative comes from (4.7) and acts on observables $\phi : [0, 1] \to \mathbb{R}$ by

$$\mathcal{L}\phi(x) = \frac{1}{4}\int_0^4 \phi(\lambda x(1 - x))\mathrm{d}\lambda - \phi(x). \tag{4.60}$$

Our goal will be to bound the expected long-time average of the state-space observable $\Phi(x) = x$ over all initial conditions $x_0 \in [0, 1]$ using the bounding techniques from Example 4.2.

To improve numerical conditioning, we let $\mathcal{D}_p = \{T_0(x), T_1, (x), \ldots, T_\alpha(x)\}$ and $\mathcal{D}_q = \{T_0(x), T_1, (x), \ldots, T_\beta(x)\}$, where $T_i(x)$ is the Chebyshev polynomial of degree i and $\beta \geq \alpha$. To implement the Chebyshev basis numerically we use the Chebfun package for MATLAB [74]. Since we are only interested in the positively invariant region $x \in [0, 1]$, we implement a localized process using the set

$$S = \{x \in \mathbb{R} |\, x(1 - x) \geq 0\}, \tag{4.61}$$

with localizing function $s(x) = x(1 - x)$. The polynomial $\sigma(x)$ that arises from this condition, as in (4.38), is also expanded in the span Chebyshev polynomials up to degree $\gamma \geq \alpha$. In practice one may increase the value of γ until results become independent of its value.

Let us first interrogate the ability of EDMD to approximate the stochastic Lie derivative in this case. Taking any $\beta \geq 2\alpha$ and $\gamma \geq \alpha - 1$ returns comparable results, and so for brevity we fix $(\beta, \gamma) = (2\alpha, \alpha - 1)$ in what follows. Now, the Koopman operator associated to the stochastic logistic map (4.59) acts on $T_1(x) = x$ by

$$\mathcal{L}T_1(x) = -\frac{1}{2} - \frac{1}{2}T_2(x) \tag{4.62}$$

and acts on $T_2(x) = 2x^2 - 1$ by

$$\mathcal{L}T_2(x) = \frac{2}{3}T_2(x) + \frac{1}{3}T_4(x), \tag{4.63}$$

where $T_4(x) = 8x^4 - 8x^2 + 1$. We compare this to the Lie derivative approximated from data using $\alpha = 2$ obtained from a single trajectory of (4.59) seeded by a random initial condition from $[0, 1]$. Using a trajectory of length $N = 10^3$ gives the matrix approximation of the Lie derivative:

$$L = \begin{bmatrix} 1.0000 & 0.0000 & 0.0000 & 0.0000 & 0.0000 \\ -0.5074 & -0.0212 & -0.5126 & -0.0291 & -0.0353 \\ 0.0047 & 0.0577 & 0.6995 & 0.0076 & 0.3626 \end{bmatrix}. \tag{4.64}$$

This matrix is the output of the EDMD process (4.55) for approximating the Lie derivative with $h = 1$. Row-wise entries of L provide the coefficients of the linear combination of elements in $\mathcal{D}_q$ used to approximate the Lie derivative on each element of $\mathcal{D}_p$. In this case the columns correspond to elements $T_0(x), T_1(x), T_2(x), T_3(x), T_4(x)$ and the rows to $T_0(x), T_1(x), T_2(x)$. Redoing the process with a trajectory of length $N = 10^6$ provides a better approximation of the true Lie derivative

$$L = \begin{bmatrix} 1.0000 & 0.0000 & 0.0000 & 0.0000 & 0.0000 \\ -0.5003 & -0.0003 & -0.5006 & -0.0011 & -0.0004 \\ 0.0004 & -0.0004 & 0.6660 & -0.0002 & 0.33256 \end{bmatrix}. \tag{4.65}$$

Table 4.2. *Upper bounds on the long-time average of the state-space observable* $\Phi(x) = x$ *for the stochastic logistic map* (4.59). *Bounds are implemented as an SDP using the framework of Example* 4.2. *The Lie derivative is estimated from* N *data points, while "Exact" bounds use the exact Lie derivative* (4.60).

	$\alpha = 2$	$\alpha = 4$	$\alpha = 6$	$\alpha = 8$	$\alpha = 10$	$\alpha = 12$	$\alpha = 14$
$N = 10^4$	0.3765	0.3162	0.3186	0.2844	0.2851	0.2858	0.2856
$N = 10^5$	0.3751	0.3126	0.3086	0.2835	0.2814	0.2775	0.2757
$N = 10^6$	0.3749	0.3124	0.3072	0.2832	0.2821	0.2758	0.2730
$N = 10^7$	0.3751	0.3126	0.3070	0.2830	0.2817	0.2766	0.2737
Exact	0.3750	0.3125	0.3069	0.2829	0.2816	0.2765	0.2736

We now see significantly improved accuracy for the approximation of the Lie derivative on elements of $\mathcal{D}_p$. Similar improvements were observed in [42], where more robust numerical results indicate that the approximation L converges to the true Lie derivative $\mathcal{L}$ at a rate of $\mathcal{O}(1/\sqrt{N})$. Although a limited demonstration, matrices in (4.64) and (4.65) appear to demonstrate this convergence since the spurious terms not represented in the true Lie derivatives (4.62) and (4.63) are on the order of $\mathcal{O}(1/\sqrt{N})$. Thus, this heuristic provides for the opportunity to improve performance by performing a SINDy-type thresholding with sparsity parameter $\lambda \sim \mathcal{O}(1/\sqrt{N})$.

Let us now turn to the numerical upper bounds on the state-space observable $\Phi(x) = T_1(x)$. These bounds are implemented as an SDP following Example 4.2 by minimizing U so that

1. $U - \Phi(x) - LV(x) - x(1-x)\sigma(x)$ is SOS;
2. $\sigma(x)$ is SOS.

Results are presented in Table 4.2 for varying amounts of data and different α, the maximal degree Chebyshev polynomial in $\mathcal{D}_p$. Also provided is the bound obtained by replacing LV with the exact Lie derivative, $\mathcal{L}V$ in (4.60). Despite the documented slow convergence of the approximate Lie derivatives L to the true Lie derivative $\mathcal{L}$ in the number of data points, the upper bounds on the expected long-time average of $\Phi(x)$ differ little over the number of data points from the value with the exact Lie derivative. From averaging long trajectories of (4.59) we can conjecture that these upper bounds are converging to 0.25, but this remains to be proven.

We can similarly bound the time average of $\Phi(x) = x$ from below using SDPs. The lower bound $\Phi(x) \geq 0$ is saturated at the fixed point $x = 0$. However, this fixed point is never included in the training data since we are gathering data along one long initial condition with $x_0 \in (0,1)$. Nonetheless, the data-driven framework provides lower bounds in the form $\overline{\Phi}(x) \geq \mathcal{O}(1/\sqrt{N})$, again aligning with the convergence of the Lie derivative. This means that the method is learning information about *all* stationary distributions despite their not being part of the training set.

4.4 ▪ Control Lyapunov Functions

In the previous section we described a method that allows for identifying Lyapunov functions from data. In this section we show how this method can be extended to control system output using a **control Lyapunov function** [257, 258]. In this case Lyapunov functions are used to determine an optimal system input to stabilize an unstable fixed point. Some of the earliest applications of SOS relaxations to dynamical systems were aimed at problems in optimal control, for which there is a rich literature on the subject [270, 120, 216, 157, 110, 39] which can be incorporated into the data-driven framework of the previous section.

We begin with a control affine system

$$\dot{x} = f(x) + g(x)u \tag{4.66}$$

for some $f, g : S \to S$ and a set $S \subseteq \mathbb{R}^d$. Assume further that there is a steady state $x_* \in S$, thus satisfying $f(x_*) = 0$, and we institute the goal of determining a control input u that drives the system (4.66) to this equilibrium. To do this, we consider a smooth auxiliary function $V : S \to \mathbb{R}$ satisfying $V(x_*) = 0$ and $V(x) > 0$ for all $x \in S \setminus \{x_*\}$. We then seek a state-dependent feedback controller $u = u_*(x)$ so that V is a Lyapunov function, meaning that

$$\mathcal{L}V := [f(x) + g(x)u_*(x)] \cdot \nabla V(x) \leq 0 \qquad \forall x \in S. \tag{4.67}$$

If such a $u_*(x)$ can be found, then this condition guarantees that controlling the system according to this state-dependent input will provide stability of the equilibrium x_*. As we saw in Example 4.1, we can further obtain *asymptotic* stability of $x = x_*$ by enforcing a strict inequality in (4.67), thereby giving that all initial conditions in S are forced to converge into x_*.

Through (4.67) we again see a Lie derivative, and so we may use the methods of the previous section to estimate its action from data. Hence, we put ourselves in the setting where the true system dynamics (4.66) are unknown, but data has been collected in the form $(X, U) \in \mathbb{R}^{d\times N} \times \mathbb{R}^{1\times N}$ and $Y \in \mathbb{R}^{d\times N}$. Data is arranged so that columns of Y correspond to columns of X exactly $h > 0$ time units in the future under the influence of the control input in the corresponding column of U. We again require dictionaries $\mathcal{D}_p$ and $\mathcal{D}_q$, now with the condition that our Lyapunov function $V \in \operatorname{span}\{\vec{p}\}$.

These data are again arranged into the matrices

$$P = \begin{bmatrix} p_1(Y_1) & p_1(Y_2) & \cdots & p_1(Y_N) \\ \vdots & \vdots & \ddots & \vdots \\ p_\ell(Y_1) & p_\ell(Y_2) & \cdots & p_\ell(Y_N) \end{bmatrix} \in \mathbb{R}^{\ell\times N} \tag{4.68}$$

and

$$Q_u = \begin{bmatrix} q_1(X_1) & q_1(X_2) & \cdots & q_1(X_N) \\ \vdots & \vdots & \ddots & \vdots \\ q_m(X_1) & q_m(X_2) & \cdots & q_m(X_N) \\ q_1(X_1)U_1 & q_1(X_2)U_2 & \cdots & q_1(X_N)U_N \\ \vdots & \vdots & \ddots & \vdots \\ q_m(X_1)U_1 & q_m(X_2)U_2 & \cdots & q_m(X_N)U_N \end{bmatrix} \in \mathbb{R}^{2m\times N}. \tag{4.69}$$

Notice that the assumption that the underlying system is control affine enters into matrix Q_u, as the first half of its rows corresponds to the estimation of the Lie derivative associated to f, while the second half is associated to $g \cdot u$. With these matrices we estimate the Koopman operator

$$K_h := PQ_u^\dagger \in \mathbb{R}^{\ell\times 2m}. \tag{4.70}$$

The Koopman approximation breaks up as $K_h = [K_h^f | K_h^g]$, with $K_h^{f,g} \in \mathbb{R}^{\ell\times m}$, representing the components coming from f and g in the control affine system (4.66). Thus, K_h acts on the control Lyapunov function $V = c \cdot \vec{p}$ by

$$K_h V := c \cdot K_h^f \vec{q} + c \cdot K_h^g \vec{q} u \tag{4.71}$$

for a control input $u = u(t) \in \mathbb{R}$. This in turn gives that the Lie derivative acts on V by

$$L_h V = \frac{K_h V - V}{h}. \tag{4.72}$$

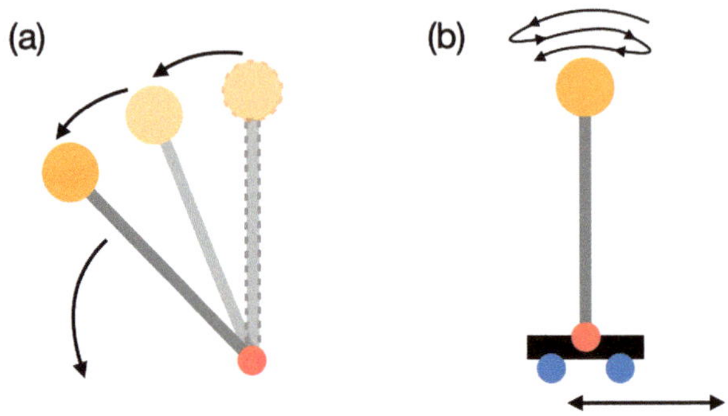

Figure 4.3. (a) *The upright position of a pendulum is unstable, meaning that slight perturbations will fall away from it.* (b) *Putting the pendulum on a cart can stabilize it in the upright position by controlling the cart's movement through small, precise jerks.*

With this approximation of the Lie derivative, we replace $\mathcal{L}V$ in (4.67) with L_hV to derive a data-driven approximation of identifying optimal controllers. Moreover, we can expand $u = u_*(x)$ as a polynomial in the state variable x with tunable coefficients so that

$$L_hV = \frac{1}{h}c \cdot (K_h^f - \Theta)\vec{q} + \frac{1}{h}c \cdot K_h^g \vec{q}u \le 0 \qquad \forall x \in S. \tag{4.73}$$

Finally, the above condition is strengthened to an SOS problem, which in turn can be solved as an SDP that identifies the coefficients of $u_*(x)$. Let us demonstrate with the following example.

Example 4.8. We consider the classical control problem of the inverted pendulum. As illustrated in Figure 4.3, when accounting for friction the hanging-down position is stable, while the standing position is unstable. Moreover, the standing position separates out how the pendulum approaches the downward position: starting to the left initially has the pendulum swinging leftward and easing into the downward position, while starting to the right initially has the pendulum swinging to the right. Now, if we put the pendulum on a wheeled cart we can introduce slight jerks of the cart to counteract the pendulum falling away from the unstable upright position—just like someone trying to balance a stack of plates in a cartoon. If we get the cart motion just right, the pendulum could be held in the unstable upright position as the cart jerks back and forth. The goal of this example will be to identify how to move the cart from data gathered from the system.[2]

Let θ denote the angle of the pendulum arm, with $\theta = 0$ being the upright position. Then the motion of the pendulum arm is captured by the second-order ODE [183]

$$\ddot{\theta} = \sin(\theta) - \varepsilon\dot{\theta} - \cos(\theta)u, \tag{4.74}$$

where $\varepsilon > 0$ is the friction coefficient and the control input u here is the acceleration of the cart on which the pendulum is placed. We fix $\varepsilon = 0.1$ throughout. Our control Lyapunov function will be taken as

$$V(\theta, \dot{\theta}) = \frac{1}{2}\dot{\theta}^2 + 1 - \cos(\theta) + \alpha(1 - \cos^3(\theta)), \quad \alpha > 0. \tag{4.75}$$

Other Lyapunov functions could be considered, and the reader is encouraged to create their own.

[2]As an aside, stabilizing the pendulum in this way has become a proxy for understanding how human beings stand using their feet as a pivot and apply small muscular adjustments to remain in the upright position. Such an understanding is becoming particularly useful in developing robots that stand on two legs, while it is being actively employed to stabilize personal transportation devices such as the Segway.

Training data is gathered by simulating 200 random initial conditions in (4.74) subject to sinusoidal external forcing of the form

$$u(t) = A\cos(t + B), \tag{4.76}$$

where A and B are drawn independently from the uniform distribution on $[-1,1]$ and $[-\pi,\pi]$, respectively. We use a timestep $h = 0.01$ and integrate each initial condition from $t = 0$ to $t = 20$. It should be noted that with a physical pendulum on a cart one can easily perform such experiments quickly and collect the resulting data for $\theta(t), \dot{\theta}(t)$ and $u(t)$. Collecting data directly from experiment could also improve control results, as models such as (4.74) are only approximate and coefficients could be estimated inaccurately, giving results that work on paper but not in experiment.

Since we are dealing with a phase variable θ which must be taken modulo 2π, we introduce the observable state variables that will be used to develop a controller:

$$x_1 = \cos(\theta), \quad x_2 = \sin(\theta), \quad x_3 = \dot{\theta}. \tag{4.77}$$

With this representation of the state variables, the Lyapunov function (4.75) becomes

$$V(x_1,x_2,x_3) = \frac{1}{2}x_3^2 + 1 - x_1 + \alpha(1 - x_1^3), \tag{4.78}$$

which is now a polynomial, and the steady state we wish to stabilize is $x_* = (1,0,0)$. The state space now requires the equality condition $x_1^2 + x_2^2 = 1$, which must be built into the SOS programs as in (4.38). Since V is cubic in x_1, we will fix the dictionary $\mathcal{D}_p$ to be all monomials in (x_1,x_2,x_3) for which x_1 and x_3 have maximal degree 3, while x_2 will have maximal degree 1. The reason we need not consider higher powers of x_2 is because the constraint $x_1^2 + x_2^2 = 1$ means that higher powers of x_2 can equivalently be written in terms of constant and x_1 terms in the dictionary. We take $\mathcal{D}_q$ to be a dictionary of all monomials with x_1, x_3 up to degree 4, while x_2 is only up to degree 1 again. Using higher powers on x_1, x_3 returns similar results.

To discover a control law $u_*(x_1,x_2,x_3)$ with an SOS program, we use the state space

$$S = \{(x_1,x_2,x_3) \in \mathbb{R} |\ \eta^2 - x_2^2 \geq 0,\ x_1^2 + x_2^2 = 1\} \tag{4.79}$$

for some parameter $\eta \in (0,1)$. The equality condition was discussed above, while the inequality condition $\eta^2 - x_2^2 \geq 0$ excludes the region where $x_1 = 0$ from the domain. The reason this is excluded from the domain is that the control input to the system, given by $\cos(\theta)u = x_1 u$, vanishes at this point, and the Lie derivative is not necessarily negative for all $x_2 = \pm 1$ and $x_3 \in \mathbb{R}$. In practice, we take η as close to 1 as possible to encapsulate as much of the full domain as we can. Numerical results presented in this example use $\eta^2 = 0.95$.

Experiments reveal that larger α causes the controlled solution to converge to the upright position $(\theta,\dot{\theta}) = (0,0)$ faster. As an example, Figure 4.4 presents controlled solutions with $\alpha = 100$. The initial condition is taken close to the hanging-down position so that the movement of the cart is forced to swing the pendulum up and over. The discovered controller to do this is

$$u_*(x_1,x_2,x_3) = \texttt{215.3461}x_1x_2 + \texttt{52.7268}x_1x_3, \tag{4.80}$$

which in terms of the original state variables is given by

$$u_*(\theta,\dot{\theta}) = \texttt{215.3461}\cos(\theta)\sin(\theta) + \texttt{52.7268}\cos(\theta)\dot{\theta}. \tag{4.81}$$

We see in Figure 4.4 that this control leads to a quick acceleration of the cart to the left and then to the right to swing the pendulum up, after which the cart hardly moves. Notice further

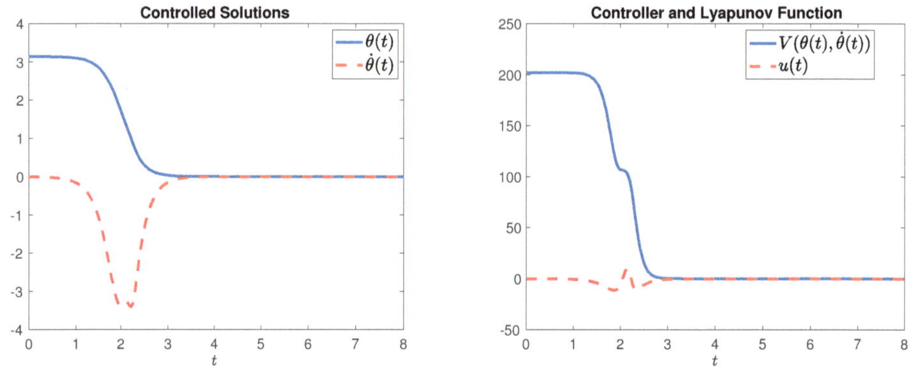

Figure 4.4. *Controlled solutions of the inverted pendulum on a cart* (4.74). *Left: The controlled arm angle and its derivative, forced into the unstable upright position at* $(\theta, \dot\theta) = (0,0)$. *Right: The monotonic approach inside the Lyapunov function* (4.75) *according to the controller,* $u(t)$, *given by* (4.81).

that the controller (4.81) vanishes at both equilibria $(\theta, \dot\theta) = (0,0), (\pi, 0)$, so it is inactive at the hanging-down ($\theta = \pi$) position. This degeneracy is likely due to the introduction of the observables (x_1, x_2, x_3) and/or the choice of Lyapunov function. Nonetheless, this is a small issue from a practical point of view, as we can randomize an initial disturbance in the pendulum using the cart, throwing it away from this equilibrium, and then turn on the controller to stabilize the system.

4.5 ▪ Extremal Invariant Measures

For many systems it is often advantageous to understand the distributional properties of typical orbits instead of focusing on individual trajectories. Referred to as statistical properties of a system, this perspective relates to the behavior of evolutions of measures. Of particular interest are measures that are preserved by the dynamics of a system. Such **invariant measures** are the primary object of study in ergodic theory and form the basis for investigation in this section. In this section we only provide the theory in a basic form and direct the interested reader to more complete introductions on measure-theoretic dynamical systems analysis, such as [47, 283, 226]. In particular, here we will review work in [41, 153], which build on earlier contributions [33, 34].

A nonnegative Borel measure μ is an invariant measure of a function $f : X \to X$ if

$$\mu(f^{-1}(A)) = \mu(A) \tag{4.82}$$

for all Borel measurable sets $A \subseteq X \subset \mathbb{R}^d$. Since we are interested in statistical properties, we only consider probability measures on X, i.e., $\mu(X) = 1$. If X is a compact invariant set of f, μ is an invariant measure if (4.82) holds everywhere along the flow forward in time. Mathematically, this can be put in terms of the Lie derivative so that μ is an invariant measure of f if and only if

$$\int_X \mathcal{L}\phi \mathrm{d}\mu = 0 \tag{4.83}$$

for all continuous observables ϕ. The existence of invariant measures for general classes of dynamical systems follows from results such as the Krylov–Bogolyubov theorem [156], and so throughout we will assume that at least one such measure exists for the system under consideration. In fact, it is typical that many invariant measures exist for a system, and it is easily shown that they form a convex subset of all probability measures on the state space X.

Since X is assumed to be compact, a measure on X is uniquely determined by its moments

$$y_k := \int_X x^k \mathrm{d}\mu, \tag{4.84}$$

where we use the multi-index notation $x^k = (x_1^{k_1}, x_2^{k_2}, \ldots, x_d^{k_d})$ for some $k \in \mathbb{N}^d$. In the absence of an underlying system, we may collect data and follow the data-driven framework of Section 4.3 to approximate the Lie derivative as $L : \operatorname{span}\{\vec{p}\} \to \operatorname{span}\{\vec{q}\}$ with $\mathcal{D}_p$ representing monomials in x up to a fixed degree $\alpha \geq 0$ and $\mathcal{D}_q$ monomials in x up to degree $\beta \geq \alpha$. Thus, for each $x^k \in \mathcal{D}_p$ the condition (4.83) is approximated by

$$\int_X L x^k \mathrm{d}\mu = 0, \tag{4.85}$$

which is a linear combination of moments of μ. The moments y_k of the measure μ are approximated by solving $Ly = 0$ for $y = [y_k]_{|k| \leq \beta} \in \mathbb{R}^m$, where $m = |\mathcal{D}_q|$ and $|k| = k_1 + k_2 + \cdots + k_d$. Notice that we do not specify whether the data is generated by a discrete versus continuous or stochastic versus deterministic process since the method applies equally to all cases by estimating L.

As we have seen previously, we want $\beta > \alpha$, guaranteeing that L has nontrivial nullspace. However, not every element of the nullspace of L leads to a measure on X. Indeed, we require that μ be a nonnegative probability measure. Enforcing that μ be a probability measure is equivalent to having the 0th moment be equal to one, i.e., $y_0 = 1$. Enforcing nonnegativity was first detailed in [119] for scalar maps, while [153] presents the following general formulation.

We again assume that X is a semialgebraic set, meaning there are polynomials g_j so that

$$X = \{x \in \mathbb{R}^d |\ g_j(x) \geq 0,\ j = 0, 1, \ldots, J\}, \tag{4.86}$$

where we define $g_0(x) = 1$ for the coming presentation. Nonnegativity of the measure on X requires that the **moment matrices** be such that $M(g_j y) \succeq 0$ for each $j = 0, 1, \ldots, J$. Taking $v_\gamma(x)$ to be the vector of all monomials in x up to degree γ, the moment matrices are defined as

$$M(g_j y) = \int_X g_j(x) v_{\gamma_j}(x) v_{\gamma_j}(x)^\intercal \mathrm{d}\mu, \tag{4.87}$$

where $\gamma_j = \lfloor (\beta - \deg(g_j))/2 \rfloor$. Therefore, putting this all together so far we have

$$\begin{aligned} Ly &= 0, \\ y_0 &= 1, \\ M(g_j y) &\succeq 0 \ \forall j - 0, 1, \ldots, J, \end{aligned} \tag{4.88}$$

which can be solved numerically as an SDP.

Best practice for implementing this method is to take β much larger than α, leading to a high-dimensional nullspace of L. This means that there are potentially numerous solutions to the problem (4.88). This plurality of solutions should be expected since, as mentioned above, systems can have more than one invariant measure. Depending on the application, we can target specific classes of measures with this approach by introducing a convex objective function $G(y)$ that we seek to minimize while satisfying (4.88). Within this framework we can target both physical and ergodic measures:

- **Physical Measures.** A *physical measure* of a transformation $f : X \to X$ is such that for Lebesgue almost every $x \in X$ and any continuous observable ϕ we have

$$\int_X \phi(x) \mathrm{d}\mu = \lim_{N \to \infty} \frac{1}{N} \sum_{n=0}^{N-1} \phi(f^n(x)). \tag{4.89}$$

Taking $\phi = x^k$, we can approximate moments of μ from sequential data gathered from the system $\{x_0, x_1, \ldots, x_N\}$ as

$$\tilde{y}_k = \frac{1}{N}\sum_{n=0}^{N-1} x_n^k \approx \int_X x^k \mathrm{d}\mu = y_k, \tag{4.90}$$

with $N \gg 1$ and all k in a user-defined set of indices $\mathcal{I}$. Thus, we optimize (4.88) with the objective to fit these moment observations by

$$G(y) = \sum_{k\in\mathcal{I}} (y_k - \tilde{y}_k)^2 \tag{4.91}$$

to approximate physical measures from data.

- **Ergodic Measures.** The extreme points of the convex set of invariant measures for a system are called *ergodic measures*. A measure is said to be ergodic if the only invariant sets of a map $f : X \to X$ have measure 0 or 1. Informally, ergodic measures provide that the domain X cannot be decomposed into nontrivial (as defined by their measure) invariant subsets of f. Ergodic measures can be targeted in (4.88) by minimizing an objective function

$$G(y) = \sum_{|k|\leq\beta} c_k y_k \tag{4.92}$$

for some $c_k \in \mathbb{R}$ that can be chosen by the user. This comes from the fact that linear objective functions in (4.88) constitute a linear program, which are known to be optimized on their extreme points, i.e., the ergodic measures.

In summary, approximating extremal measures (optima of the objective function) is formulated as the SDP

$$\min_{y\in\mathbb{R}^m} G(y) \quad \text{such that} \quad \begin{cases} Ly = 0, \\ y_0 = 1, \\ M(g_j y) \succeq 0, \quad j = 0, 1, \ldots, J. \end{cases} \tag{4.93}$$

With an optimizer y_* of the above SDP we define a polynomial of degree α in x given by

$$\rho_\alpha = M_L^{-1} y_* \cdot v_\alpha(x), \tag{4.94}$$

where $M_L \in \mathbb{R}^{m\times m}$ is the moment matrix of the Lebesgue measure on X of degree 2α:

$$M_L = \int_X v_\alpha(x) v_\alpha(x)^\intercal \mathrm{d}x. \tag{4.95}$$

The polynomial ρ defines a Lebesgue-absolutely-continuous measure

$$\mu_\alpha(A) = \int_A \rho_\alpha(x)\mathrm{d}x \qquad \forall A \subseteq X, \tag{4.96}$$

which approximates an extremal invariant probability measure of the underlying system.

Convergence properties. Notice that μ_α is a signed measure since ρ need not be nonnegative throughout X. Furthermore, μ_α approximates μ in the weak-* topology, meaning

$$\int_X g(x)\rho_\alpha(x)\mathrm{d}x \approx \int_X g(x)\mathrm{d}\mu \qquad \forall g : X \to \mathbb{R}. \tag{4.97}$$

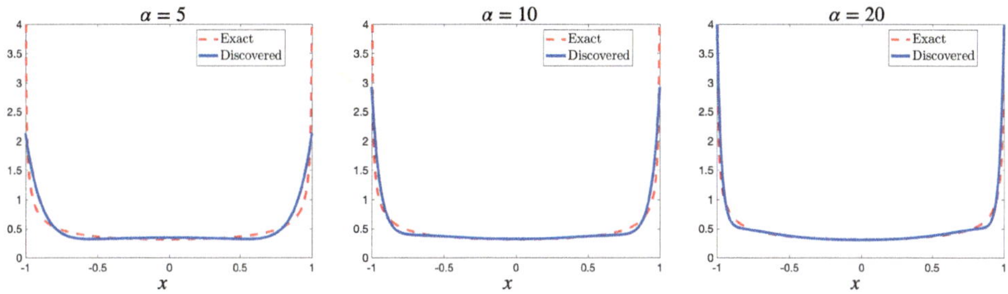

Figure 4.5. *Discovered approximations of the physical measure density* (4.99) *for the mapping* (4.98) *via the SDP* (4.93) *using* $N = 10^4$ *data points. Here* $\alpha = 5, 10, 20$ *is the maximal degree of the Chebyshev polynomial dictionary* $\mathcal{D}_p$, *while the maximal degree of* $\mathcal{D}_q$ *is* 2α.

Thus, notions of convergence in the amount of data or in the polynomial degree α should not be stated in terms of ρ_α but in terms of statistical information coming from integrating with respect to these measures. More precisely, if we use the exact Lie derivative in (4.93) and a polynomial map f, [153] proves under mild conditions that the measures μ_α converge to an invariant measure μ in this weak-* topology. These results were extended to the data-driven setting (4.93) in [41].

Example 4.9. We consider data generated by the deterministic quadratic mapping

$$x_{n+1} = 2x_n^2 - 1, \tag{4.98}$$

which has the invariant semialgebraic domain $X = [-1, 1] = \{x \in \mathbb{R} : 1 - x^2 \geq 0\}$. Moreover, this map is known to have a physical measure with density given by [168]

$$\rho_*(x) = \frac{1}{\pi\sqrt{1-x^2}}, \tag{4.99}$$

which we can compare with the densities (4.94) that are identified from data generated by (4.98). Localizing to the domain X means $g_1(x) = 1 - x^2$, and (4.93) has two moment matrices.

We approximate the physical measure from data collected by iterating (4.98) from the initial condition $x_0 = 0.25$. Our objective will be to fit the first moment approximated from data using

$$G(y) = (y_1 - \tilde{y}_1)^2, \tag{4.100}$$

where $\tilde{y}_1$ is the average of the provided data, according to (4.90). Dictionaries are taken to be $\mathcal{D}_p = \{T_0(x), T_1, (x), \ldots, T_\alpha(x)\}$ and $\mathcal{D}_q = \{T_0(x), T_1, (x), \ldots, T_\beta(x)\}$, where we again use a Chebyshev basis for our SDPs to improve numerical conditioning. Taking $\beta \geq 2\alpha$ returns nearly identical results, and so we fix $\beta = 2\alpha$ throughout these demonstrations.

Approximations of the density using $N = 10^4$ data points are plotted in Figure 4.5 for $\alpha = 5, 10, 20$. While we can see the approximated densities starting to line up with the true density (4.99) as α increases, noticeable oscillations are also forming. This is not necessarily an issue though since convergence of measures does not come in the form of pointwise convergence of densities but in the weak-* topology described in (4.97). In this case, weak-* convergence is equivalent to convergence in the L^1-norm of the cumulative density functions [41]

$$R_\alpha(x) = \int_{-1}^{x} \rho_\alpha(s)\mathrm{d}s, \qquad R_*(x) = \int_{-1}^{x} \rho_*(s)\mathrm{d}s = \frac{1}{2} + \frac{1}{\pi}\arcsin(x), \tag{4.101}$$

which we compare for different amounts of data N and degrees α in Table 4.3. Here we see that

Table 4.3. *The L^1-norm error between the discovered and exact cumulative distributions* (4.101). *The Lie derivative is identified using N data points, and the SDP* (4.93) *has the objective to fit the approximate first moment $\tilde{y}_1$ from the data. Exact values are computed using the exact Lie derivative and the true first moment $y_1 = 0$ in place of the data-driven estimations.*

	$\alpha = 5$	$\alpha = 10$	$\alpha = 15$	$\alpha = 20$	$\alpha = 25$	$\tilde{y}_1$
$N = 10^2$	0.04488	0.04401	0.04376	0.04336	0.04325	-0.04123
$N = 10^3$	0.02717	0.01602	0.01474	0.01440	0.01389	-0.01339
$N = 10^4$	0.02872	0.00955	0.00531	0.00372	0.00285	-0.00102
$N = 10^5$	0.02891	0.00959	0.00545	0.00371	0.00278	-0.00035
Exact	0.02902	0.00962	0.00556	0.00378	0.00280	0.0000

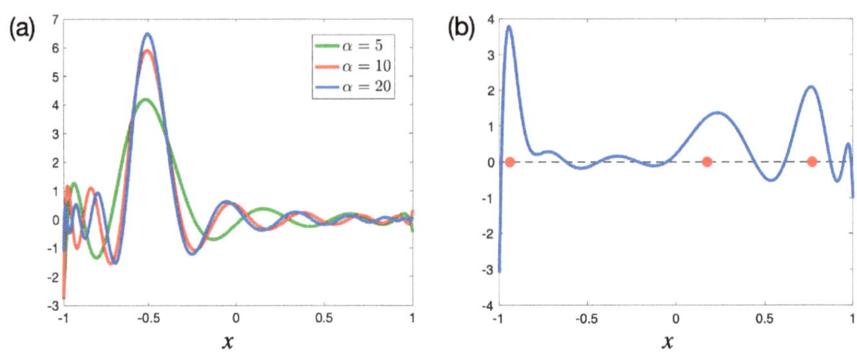

Figure 4.6. (a) *Discovered densities of signed measures that become increasingly localized about the fixed point $x = -0.5$ of* (4.98) *by increasing the maximal polynomial degree α of the dictionary $\mathcal{D}_p$.* (b) *A discovered density of a signed measure with $\alpha = 10$ that appears to be localized about a period-3 orbit at the points $x = -0.93969, 0.76604, 0.17365$, indicated by the red dots along the x-axis.*

results are comparable to implementing the method with the exact Lie derivative (bottom row), showing that the main limitation is approximating ρ_* using polynomials rather than approximating the Lie derivative from data, likely because of the singularities in ρ_* at $x = \pm 1$.

Using a linear objective function allows for the approximation of ergodic measures from data. For example, estimating the Lie derivative with $N = 10^3$ data points and minimizing $G(y) = y_1$ results in densities of signed measures that become increasingly localized around $x = -0.5$, as shown in Figure 4.6. This point $x = -0.5$ is a fixed point of the map (4.98), meaning we are approximating an **atomic measure**. Atomic measures are linear combinations of delta functions supported on the cyclic points of the map. Importantly, we can extract the approximate atoms from the computed signed measures as the minimizers of the dual problem in polynomial optimization [171]. We refrain from providing the details of this extraction here but note that the YALMIP fork `aeroimperial-yalmip`[3] contains the function `extractminimizers` to provide this information. In this case we find a single atom of the measure at $x = -0.5$, as expected.

One may continue to use different linear objective functions to identify different approximations of atomic measures from data. Moreover, these atoms always lie along cyclic orbits of the map (4.98), thus providing a method of extracting periodic orbits directly from data. For example, using instead the optimization objective $G(y) = y_3$ one identifies the density plotted

[3]Available at https://github.com/aeroimperial-optimization/aeroimperial-yalmip.

in Figure 4.6(b) with minimizers lying along a 3-cycle of (4.98), given by

$$x = -\texttt{0.93969}, -\texttt{0.17365}, \texttt{0.76604}. \tag{4.102}$$

These points are indicated by red dots along the x-axis in Figure 4.6. One should be reminded of the fact that none of these orbits are included in the training data, and therefore this structure is entirely identified via the approximate Lie derivative and moment optimization problem (4.88).

In general, there is no way of predicting which linear objective functions identify which cyclic points of the map. For example, using $G(y) = y_7$ returns a density with minimizers at a 6-cycle and the fixed point $x = -0.5$. However, while determining which atoms will be present for a given optimization objective is not possible, in practice one may randomize linear objective functions $G(y)$ to identify various cyclic points. Since these SDPs can be solved quickly, sweeping over numerous optimization objectives can be done with relative ease.

Approximating invariant measures from data is an old problem in dynamical systems that goes back at least to Ulam [276]. There are many competing methods to do this that were not reviewed here, including Ulam's method, constructing histograms, fitting Gaussian mixtures, kernel density estimation, and variants of EDMD [72, 105, 147, 148, 94]. In [41] the SDP approach is shown to outperform histograms and Gaussian mixtures on the examples therein, while benefits over other methods are also highlighted. The main advantage of the SDP approach is that it can identify more than one invariant measure from the same data, including atomic measures associated with unstable trajectories. Other methods can only discover the particular invariant measure from which the data is sampled—typically the physical one.

Finally, discussions of invariant measures for dynamical systems typically focus on the **Perron–Frobenius operator**. The exposition here has purposely skirted the introduction of this operator to be consistent with the Koopman operator approach, which was the theme of this chapter. For completion, the Perron–Frobenius operator is a linear operator $\mathcal{P}$ associated to a function $f : X \to X$ that acts on the space of measures by

$$(\mathcal{P}\mu)(A) = \mu(f^{-1}A), \qquad A \subseteq X. \tag{4.103}$$

Notice from (4.82) that invariant measures of f are exactly fixed points of $\mathcal{P}$. In fact, the Perron–Frobenius operator is known to be dual to the Koopman operator, making them in some sense two sides of the same coin. Furthermore, many of the competing methods for identifying invariant measures from data take the dual perspective of the Perron–Frobenius operator, including Ulam's method [276] and other EDMD approaches [147, 148].

4.6 ▪ Applications of Auxiliary Functions and Beyond

Throughout this chapter we have seen applications of polynomial optimization and SDPs to obtain system-level information from data. This was primarily done through using auxiliary functions but also included statistical measures. Further applications of auxiliary functions to bounding time averages and invariant measures can be found in [58, 83, 184, 31, 103]. Beyond the applications provided in this section, auxiliary functions have been used to approximate reachable sets [151, 185], basins of attraction [227, 217], and attractors [134, 245, 246]. They can further be employed to estimate system parameters and propagate uncertainty [263, 262, 196, 64] and solve optimal control and optimal stopping problems [123, 59, 170, 122]. Importantly, all of these applications encode system dynamics through the Lie derivative and are therefore amenable to the data-driven framework outlined in this section. Moreover, innovative methods are being developed to push these methods into the analysis of PDEs [57, 82, 121, 277, 152, 84], meaning that there is a possibility to extend such data-driven analysis to spatiotemporal systems.

This chapter primarily provided a targeted application of EDMD. The result is another dictionary-based method for interpreting dynamic data. Like any dictionary-based algorithm, as the dimension of the underlying data grows, the problem becomes intractable due to an explosion in the size of the dictionary required for good performance. One method to counter this is to enforce sparsity in the SOS problem [297], but this cannot be enforced without knowledge of a sparsity pattern. Therefore, like SINDy and EDMD, the methods outlined in this section are primarily applicable to low-dimensional data. Extensions to high-dimensional data, potentially using kernel formulations as in Section 2.5, are thus emerging frontiers in polynomial optimization.

Chapter 5

Learning Dynamics with Neural Networks

In this chapter we turn our study to neural networks. As their name suggests, neural networks were originally posed to model neurons in the brain. The origins of neural networks are typically traced back to 1943 with the work of McCulloch and Pitts, who built an electrical circuit that attempted to model a network of neurons [192]. Neural networks have come a long way from this foundational work, with their mathematical structure being influenced by advancements in neuroscience throughout the twentieth century; see [182] for a history. Although neural networks were originally restricted to supercomputers in private labs, the growth in personal computing power now means that almost anyone can implement a neural network from the comfort of their own home to perform a number of tasks. As will be emphasized in this chapter and the next, the widespread availability of open-access software to build and train neural networks has been particularly promising for forecasting, discovering, and interpreting the behavior of dynamic systems.

In this chapter we introduce the application of neural networks to dynamical systems analysis. We begin by introducing their mathematical theory and then move through a number of applications and recent advancements. Most of our examples in this chapter will use the open-source software TensorFlow 2.0 [1] to build and train the network, necessitating only the moderate level of computing power that most modern personal computers possess to train them.

5.1 ▪ Universal Function Approximators

In the previous chapters we saw that we can fit data to curves by performing a dictionary expansion and regressing the data onto the dictionary by appropriately tuning coefficients in the linear span. In all such situations the unknowns, i.e., the coefficients, appear linearly and thus lead to finding a matrix that minimizes the error. Although this process is easy to implement and includes explicit results giving the existence of minimizing matrices, results are highly dependent on the dictionaries used to approximate the curves. Neural networks provide an alternative method of curve fitting that does not rely on dictionary expansions but uses classes of functions that can be manipulated through a training process to approximate any function to an arbitrary accuracy. Let us begin with an example that illustrates a class of these functions and how they can approximate any continuous function on an interval.

Example 5.1. We begin with an arbitrary continuous function $f : [0,1] \to \mathbb{R}$. Since the domain of f is the compact interval $[0,1]$, f is uniformly continuous, which as a consequence implies

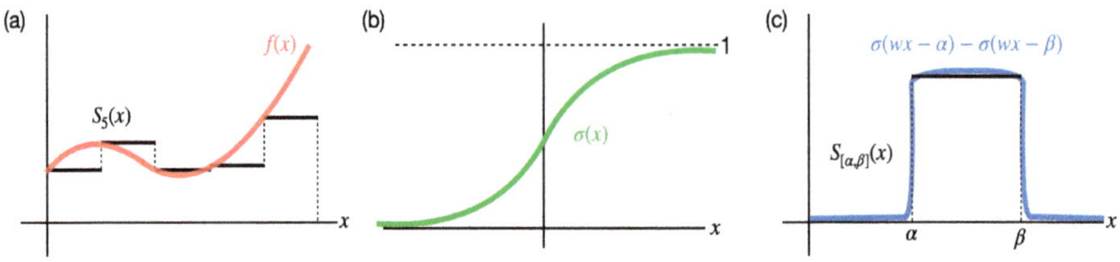

Figure 5.1. (a) *Approximation of a continuous function f by step functions.* (b) *A sigmoid function.* (c) *Approximation of a step function by a linear combination of scaled sigmoids.*

that for any $\varepsilon > 0$ we can find an $n \geq 1$ so that the piecewise constant (step) function

$$S_n(x) = \begin{cases} f(0), & x \in [0, 1/n), \\ f(1/n), & x \in [1/n, 2/n), \\ \vdots & \\ f(1-1/n), & x \in [1-1/n, 1], \end{cases} \tag{5.1}$$

satisfies

$$\|f - S_n\|_\infty := \sup_{x \in [0,1]} |f(x) - S_n(x)| < \varepsilon. \tag{5.2}$$

Here the function $S_n(x)$ assigns the constant value $f(k/n)$ to every x in the interval $[k/n, (k+1)/n)$, and uniform continuity of f guarantees that taking n large enough approximates f to any accuracy. See Figure 5.1(a) for an illustration using $n = 5$. The drawback is that step functions are not continuous, and so we have approximated a continuous function by one that is not.

One way to approximate f by a class of continuous functions is to approximate step functions by continuous functions so that through transitivity (the triangle inequality) one can in turn approximate f. This is the direction that was taken by Cybenko's 1989 work, which provided a class of continuous functions that can approximate any other continuous function [68]. Cybenko used **sigmoid functions**, S-shaped monotone increasing functions $\sigma(x)$ with limits

$$\lim_{x \to -\infty} \sigma(x) = 0, \qquad \lim_{x \to \infty} \sigma(x) = 1, \tag{5.3}$$

as illustrated in Figure 5.1(b). Examples include the logistic function

$$\sigma(x) = \frac{1}{1 + \mathrm{e}^{-x}} \tag{5.4}$$

and the (shifted) arctangent

$$\sigma(x) = \frac{1}{2}\arctan(x) + \frac{1}{2}, \tag{5.5}$$

among many others.

To see how sigmoid functions approximate step functions, let us begin with an arbitrary subinterval $[\alpha, \beta] \subset [0, 1]$ and a step function with a single step,

$$S_{[\alpha,\beta]} = \begin{cases} 1, & x \in [\alpha, \beta], \\ 0, & x \notin [\alpha, \beta]. \end{cases} \tag{5.6}$$

Notice that the step functions S_n in (5.1) can be written as the linear combination of step functions $S_{[\alpha,\beta]}$ for various $\alpha, \beta \in [0, 1]$ by

$$S_n(x) = \sum_{k=1}^{n} f((k-1)/n) S_{[(k-1)/n, k/n]}(x), \tag{5.7}$$

and so approximating an arbitrary $S_{[\alpha,\beta]}$ with sigmoid functions can be built up to $S_n(x)$ by approximating each $S_{[(k-1)/n,k/n]}$ and summing these approximations over k.

With a scaling coefficient inside the sigmoid function, $\sigma(wx)$, the limits (5.3) give

$$\lim_{w\to\infty} \sigma(wx) = \begin{cases} 1, & x > 0, \\ 0, & x < 0, \end{cases} \tag{5.8}$$

pointwise in x. So, taking $w \gg 1$ allows one to define a smooth, sharp transition from 0 to 1 over the real line. We can move the location of the transition by shifting it to $x = \alpha$, the location of the step transition from 0 to 1 in $S_{[\alpha,\beta]}$, resulting in $\sigma(wx - \alpha)$ with $w \gg 1$. So, we now have a smooth transition from 0 to 1 around $x = \alpha$ but no transition back down at $x = \beta$. To remedy this we notice that $-\sigma(wx - \beta)$ has a sharp transition down from 0 to -1 at $x = \beta$ when $w \gg 1$, and so it can be used to cancel the ongoing contribution of $\sigma(wx - \alpha)$ beyond $x \geq \beta$. This results in the approximation

$$S_{[\alpha,\beta]}(x) \approx \sigma(wx - \alpha) - \sigma(wx - \beta), \quad w \gg 1, \tag{5.9}$$

thus giving a smooth approximation to a step function using sigmoid functions, as illustrated in Figure 5.1(c). Doing this over the subintervals $[k/n, (k+1)/n]$ gives the approximation of S_n,

$$S_n(x) \approx \sum_{k=1}^{n} f(k/n)[\sigma(w_k x - (k-1)/n) - \sigma(w_k x - k/n)], \tag{5.10}$$

which in turn provides a smooth approximation of f that can be made more accurate by increasing n and the w_k. Notice that the only functions that appear in the approximation of S_n are sigmoid functions, albeit ones that are scaled and shifted. This indicates that we can approximate continuous functions using only linear combinations of scaled sigmoid functions.

The previous example is not meant to be rigorous but only to provide some intuition for Cybenko's much more general result that says that any continuous function $f : [0,1]^M \to \mathbb{R}$ can be approximated to arbitrary accuracy in the $\|\cdot\|_\infty$ norm by functions of the form

$$g(x) = \sum_{k=1}^{n} v_k \sigma(w_k \cdot x + b_k), \tag{5.11}$$

where $n \geq 1$, $w_k \in \mathbb{R}^M$, $v_k, b_k \in \mathbb{R}$, and σ is a sigmoid function [68]. What this means is that the set of functions of the form (5.11) is dense in the space of continuous functions from $[0,1]^M$ to $\mathbb{R}$. Cybenko's result is the first of a number of results called **universal approximation theorems** that provide the theoretical underpinning of the approximation capabilities of neural networks.

There have been many improvements to Cybenko's result. We now know that for any continuous nonpolynomial $\sigma : \mathbb{R} \to \mathbb{R}$ and a continuous function $f : C \to \mathbb{R}^N$, where $C \subset \mathbb{R}^M$ is compact, we can approximate f to arbitrary accuracy in the $\|\cdot\|_\infty$ norm by functions of the form

$$g(x) = V \cdot \sigma(Wx + b), \tag{5.12}$$

where $W \in \mathbb{R}^{n\times M}$, $V \in \mathbb{R}^{N\times n}$, $b \in \mathbb{R}^n$, and σ is understood to act by $[\sigma(x)]_m = \sigma(x_m)$ [223, 127, 256, 115]. It should be emphasized that we no longer require sigmoids to perform our approximations, but any nonpolynomial function will suffice. Any such function σ is referred to as an **activation function** in the context of neural networks.

In (5.12) there is a "hidden" dimension $n \geq 1$ which comes from $Wx + b \in \mathbb{R}^n$ inside σ. This value is referred to as the **width**, and universal approximation theorems rely on taking n large to get more accurate function approximations. As one might further expect, some functions may require a very large value of n to be approximated within a desired accuracy and therefore may require a large number of computational resources. Sometimes we can overcome this computational bottleneck by considering instead functions of the form

$$g(x) = W_d \cdot \sigma(W_{d-1} \cdot \sigma(W_{d-2} \cdots (W_2 \cdot \sigma(W_1 x + b_1) + b_2) + \cdots) + b_{d-1}) + b_d, \quad (5.13)$$

taken to be d compositions of functions of the form (5.12). The parameter d is referred to as the **depth** and represents the number of compositions that are taken to form the function. Note further that inside each activation function there may be different widths, n_j with $j = 1, \ldots, d$. There are universal approximation theorems that apply to functions of the form (5.13) to deal with the case of bounded widths, but the depth can be as large as necessary. The work [179] shows that if we fix all widths to be $N + 4$ and take σ as the ReLU activation function,

$$\mathrm{ReLU}(x) = \max\{0, x\}, \quad (5.14)$$

then functions of the form (5.13) can approximate any Lebesgue integrable function to arbitrary accuracy in the L^1-norm. Similar results have been established for approximating continuous functions with bounded widths and arbitrary depths, such as [144], which allows for any non-affine continuous activation function σ. There are many generalizations of these results that are tailored to specific function spaces or tasks in machine learning, but we do not review them here. We have only provided the basics to justify the performance of the coming computational examples.

5.2 ▪ Neural Networks

The theory in the previous section provides the foundation for modern neural networks. A **single-layer neural network** is styled upon (5.12) and given by a function with input $x \in \mathbb{R}^M$ and output $y \in \mathbb{R}^N$ so that

$$y = \sigma(Wx + b) \quad (5.15)$$

for an activation function $\sigma : \mathbb{R} \to \mathbb{R}$, $W \in \mathbb{R}^{N \times M}$, and $b \in \mathbb{R}^N$. The entries of the matrix W are the **weights**, and the vector b is the **bias** of the neural network, both of which are tunable to adjust the output of the network. The function $\sigma(Wx + b)$ constitutes a **layer** of the network.

Comparing (5.15) and the universal approximators (5.12), our single-layer neural network is not general enough to guarantee that it can approximate any function, and so we may generalize to neural networks of greater depth by considering the output y given by

$$y = \sigma_d(W_d \cdot \sigma_{d-1}(W_{d-1}(\cdots (W_2 \cdot \sigma_1(W_1 x + b_1) + b_2) + \cdots) + b_{d-1}) + b_d), \quad (5.16)$$

which is akin to (5.13) and results from composing layers of the neural network with each other. Here each $W_j \in \mathbb{R}^{n_{j+1} \times n_j}$ and $b_j \in \mathbb{R}^{n_{j+1}}$ with some $n_j \geq 1$ for each $j = 1, \ldots, d$ and $n_d = N$ fixed. The activation for each layer may be the same or may be different. For example, if $d = 2$ with σ_1 a sigmoid function and σ_2 the identity function, then we arrive at functions in the form of Cybenko's universal approximators from (5.11). When d is large, the neural network (5.16) is said to be **deep**, while if some n_j is large, then the jth layer is said to be **wide**. The output of the jth network layer becomes the input for the $(j + 1)$th layer, creating a feedforward structure taking the input from layer to layer to produce the output, as illustrated in Figure 5.2.

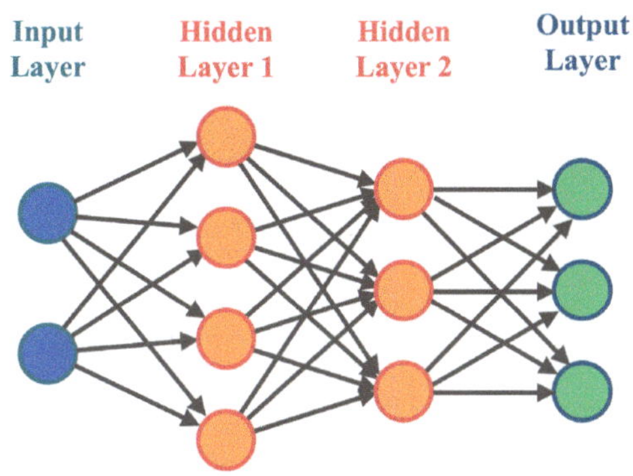

Figure 5.2. *An illustration of the feedforward neural network architecture according to* (5.16). *In this caricature there are two hidden layers. The input is passed to the first hidden layer by applying* $\sigma_1(W_1x + b_1)$*, which is then used as the input for* $\sigma_2(W_2x + b_2)$ *to arrive at the second layer, which then arrives at the output by applying* $\sigma_3(W_3x + b_3)$ *to the output at the second layer. Here the depth is* 3*, and the widths are* 4 *for the first hidden layer and* 3 *for the second.*

Now, suppose we are given data in the form of matrices

$$X = \begin{bmatrix} | & | & & | \\ X_1 & X_2 & \cdots & X_K \\ | & | & & | \end{bmatrix} \in \mathbb{R}^{M \times K}, \qquad Y = \begin{bmatrix} | & | & & | \\ Y_1 & Y_2 & \cdots & Y_K \\ | & | & & | \end{bmatrix} \in \mathbb{R}^{N \times K} \tag{5.17}$$

and we wish to identify a function that continuously maps data points $X_k \mapsto Y_k$ for each $k = 1, \dots, K$. Using the theory of universal function approximations, if such a function existed, then we should be able to approximate it by (5.16) with appropriately chosen $W_j \in \mathbb{R}^{n_{j+1} \times n_j}$, $b \in \mathbb{R}^{n_{j+1}}$, $n_j \geq 1$, and $d \geq 1$. Let us fix d and the n_j, and let $g(x)$ denote the neural network (5.16), so that we want to fit $Y_k \approx g(X_k)$ over the data points by identifying appropriate weights W_j and biases b_j for $j = 1, \dots, d$. To quantify the accuracy of this data fitting, we consider the **mean-square error** (mse) between $g(X_k)$ and Y_k, leading to the smooth loss function

$$\mathcal{L} = \|Y - g(X)\|_{\text{mse}} := \frac{1}{2K} \sum_{k=1}^{K} \|Y_k - g(X_k)\|_2^2. \tag{5.18}$$

Recall that $\|Y_k - g(X_k)\|_2$ represents the standard Euclidean norm for vectors in $\mathbb{R}^M$, and note that the loss $\mathcal{L}$ depends on the weight and bias parameters of the neural network. The goal is to minimize the nonlinear loss function $\mathcal{L}$ over the possible neural network parameters so that we bring the network output $g(X_k)$ as close as possible to the desired output Y_k.

Training the neural network is performed by minimizing the loss function (5.18) through an iterative process called **gradient descent**, which is summarized in Algorithm 5.1. Initializing the network parameters randomly, we then define a parameter $\gamma > 0$, the **learning rate**, to produce the iterative updates

$$\begin{aligned} W_j^{\text{new}} &= W_j^{\text{old}} - \gamma \frac{\partial \mathcal{L}}{\partial W_j}, \\ b_j^{\text{new}} &= b_j^{\text{old}} - \gamma \frac{\partial \mathcal{L}}{\partial b_j} \end{aligned} \tag{5.19}$$

for each $j = 1, \dots, d$. The gradients $\partial \mathcal{L}/\partial W_j \in \mathbb{R}^{n_{j+1} \times n_j}$ and $\partial \mathcal{L}/\partial b_j \in \mathbb{R}^{n_{j+1}}$ are evaluated at all of the old parameter values, and we repeat these updates by replacing the "new" values

Algorithm 5.1 Gradient Descent

1. Initialize network weights W and biases b. Define loss function $\mathcal{L}$ to minimize
2. Fix learning rate $\gamma > 0$ and loop until convergence:
   (a) Compute partial derivatives $\partial\mathcal{L}/\partial W$ and $\partial\mathcal{L}/\partial b$ at current weight W and bias b
   (b) Update $W \leftarrow W - \gamma\partial\mathcal{L}/\partial W$ and $b \leftarrow b - \gamma\partial\mathcal{L}/\partial b$
   (c) Use updated W and b as current weights and biases and return to (a)
3. Converged W and b constitute trained neural network

with the "old" ones at each step to gradually decrease the value of the loss $\mathcal{L}$. The learning rate γ represents the size of the step from the old values to the new ones. Larger γ leads to a faster training process, but if taken too large, then iterations will fail to converge to a local minimum of the loss function because they will keep jumping over it.

If $d \geq 2$, then computing the gradients $\partial\mathcal{L}/\partial W_j$ and $\partial\mathcal{L}/\partial b_j$ can be complicated due to the function compositions in the network. For example, for the weight matrices W_j we have

$$\frac{\partial\mathcal{L}}{\partial W_j} = -\frac{1}{K}\sum_{k=1}^{K}\sum_{m=1}^{M}([Y_k]_m - [g(X_k)]_m)\frac{\partial g}{\partial W_j}, \qquad j \in \{1,\ldots,d\}, \tag{5.20}$$

where $[\cdot]_m$ denotes the mth component of the vector. With $d = 2$, the chain rule gives

$$\frac{\partial g}{\partial W_1} = \sigma_2'(W_2\tilde{y} + b_2) * W_2 \circ \sigma_1'(W_1x + b_1) \otimes x^\intercal, \tag{5.21}$$

where the notation $\sigma'(W_2\tilde{y} + b_2) * W_2$ represents each row of $\sigma'(W_2\tilde{y} + b_2)$ multiplied against every element of the same row in W_2, $\circ$ is the Hadamard product, and $\otimes$ is the Kronecker product.[4] Similarly, the partial derivative with respect to W_2 gives

$$\frac{\partial g}{\partial W_2} = \sigma_2'(W_2\tilde{y} + b_2) \otimes \tilde{y}^\intercal. \tag{5.22}$$

Although the notation is complicated, we see that both gradients have the term $\sigma_2'(W_2\tilde{y} + b_2)$. Hence, we can compute $\partial g/\partial W_2$ first and store this common value to use it in the computation of $\partial g/\partial W_1$. The result is a reduction in computational complexity since information from the last layer is being pushed backward through the network to speed up computations in the hidden layer. Furthermore, returning to the loss gradient (5.21) one has

$$\frac{\partial\mathcal{L}}{\partial W_2} = -\frac{1}{K}\sum_{k=1}^{K}[(Y_k - g(X_k)) \circ \sigma_2'(W_2\tilde{y} + b_2)] \otimes \tilde{y}^\intercal \tag{5.23}$$

and

$$\frac{\partial\mathcal{L}}{\partial W_1} = -\frac{1}{K}\sum_{k=1}^{K}[(Y_k - g(X_k)) \circ \sigma_2'(W_2\tilde{y} + b_2)] * W_2 \circ \sigma_1'(W_1x + b_1) \otimes x^\intercal, \tag{5.24}$$

[4]The notation is complicated here because of the convention that $\sigma(Wx + b)$ denotes the function σ acting on each row of the vector $Wx + b$.

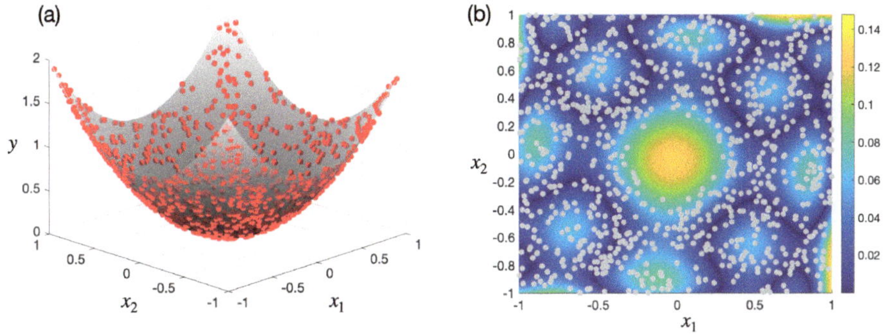

Figure 5.3. (a) *Training data (red) sampled randomly from the function $y = x_1^2 + x_2^2$ (gray).* (b) *A contour plot of the difference in absolute value between the true function $y = x_1^2 + x_2^2$ and the trained neural network with the training points overlaid (gray dots).*

and so we can pass the computation of $(Y_k - g(X_k)) \circ \sigma_2'(W_2\tilde{y} + b_2)$ for each k back to the gradient $\partial\mathcal{L}/\partial W_1$, again reducing the number of computations needed to compute the gradients.

Deeper networks allow for more information to be pushed backward through the network from each gradient computation. In particular, if $\tilde{y}_j$ is the output of the jth layer of the neural network, only $\sigma_j'(W_j\tilde{y}_{j-1} + b_j)$ needs to be computed for the gradient $\partial g/\partial W_j$. This method of updating gradients from the last layer backward through the network is called **backpropagation** [104, 238] and offers a quick and efficient way to train a neural network through gradient descent.

Example 5.2. Let us now demonstrate training a neural network to approximate the function $f(x_1, x_2) = x_1^2 + x_2^2$ from data. Input data $X \in \mathbb{R}^{2\times 1000}$ comes from drawing $K = 1000$ pairs (x_1, x_2) from the uniform distribution on $[-1, 1]^2$, and output data $Y \in \mathbb{R}^{1\times 1000}$ comes from the evaluation of each pair in X by f. A realization of these training points is plotted in Figure 5.3(a). We will employ a network architecture in the form of Cybenko's universal approximation theorem, i.e., $x \mapsto W_2\sigma(W_1x+b)$, with a logistic activation function (5.4). Recall that we have seen the logistic function before in Example 3.8, and we know that it satisfies the differential equation

$$\sigma'(x) = \sigma(x)(1 - \sigma(x)). \tag{5.25}$$

This means that the derivatives in the gradient descent update can be computed with fewer operations using $\sigma(W_1x + b_1)$, which will have already been computed.

Our demonstration uses a hidden layer with width $n = 100$, meaning $W_1 \in \mathbb{R}^{100\times 2}$, $b \in \mathbb{R}^{100}$, and $W_2 \in \mathbb{R}^{1\times 100}$, giving a total of

$$\underbrace{2 \times 100}_{W_1} + \underbrace{100}_{b_1} + \underbrace{100}_{W_2} = 400 \tag{5.26}$$

trainable network parameters. The learning rate will be $\gamma = 0.1$, and we perform 10^5 iterations of gradient descent from randomly initiated W_1, W_2, and b on the loss function (5.18). In practice one should repeat the training process with different hidden widths, learning rates, and even depths of the neural network to identify the network architecture which gives the smallest loss.

The loss is initiated as $\mathcal{L} =$ `741.3845` with random network parameters. After 10^5 parameter updates with gradient descent, the loss is decreased to $\mathcal{L} =$ `0.0030`. Figure 5.3(b) provides a contour plot of the pointwise error $|f(x_1, x_2) - g(x_1, x_2)|$ over $[-1, 1]^2$, where g is the trained neural network. The training data points are overlaid as gray dots, and we notice that the largest error is near $(x_1, x_2) = (0, 0)$, where there is an absence of training data. This should be expected since the network has no knowledge of what it is approximating here and so has nothing to guide its output. Notice further that in regions where sampling is high, the error is low, which should again be expected since the network has a lot of information for interpolating in these regions.

5.3 ▪ Forecasting Dynamics

In the previous section the neural network was constructed by hand, and the loss was reduced through tedious gradient computations. Fortunately, there are now free and open-source software libraries that can automate building and training a neural network. Notable examples include TensorFlow [1] and PyTorch [219], of which we will use the former for the remaining experiments in this text. These software libraries include optimized backpropagation routines, stochastic gradient descent [36, 145, 13], and numerous activation functions to choose from. One advantage we gain here is that we can introduce more complicated loss functions and neural network architectures that allow for the approach to a diversity of problems in dynamical systems without having to build the network from scratch or be bogged down in the details of training it.

In this section we provide a variation of the standard loss function (5.18) to incorporate dynamical information from the data. We consider sequential data arranged into the matrix

$$X = \begin{bmatrix} | & | & & | \\ X_1 & X_2 & \cdots & X_K \\ | & | & & | \end{bmatrix} \in \mathbb{R}^{M \times K}. \tag{5.27}$$

Since the data is sequential, we wish to learn a mapping that best approximates iterating from one instance to the next, i.e., learning the correspondence $X_k \mapsto X_{k+1}$ for all k. Letting g represent our neural network, we can use the loss function from the previous section (5.18), given by

$$\mathcal{L} = \frac{1}{2(K-1)} \sum_{k=1}^{K-1} \|X_{k+1} - g(X_k)\|_2^2. \tag{5.28}$$

Indeed, such a loss function aims to have $g(X_k) \approx X_{k+1}$ for all $k = 1, \ldots, K-1$. However, we can improve the loss function by incorporating knowledge that the data is sequential. That is, $g^2(X_k) := g(g(X_k))$ should approximately be X_{k+2}, and in general

$$g^s(X_k) := \underbrace{g \circ \cdots \circ g}_{s \text{ times}}(X_k) \approx X_{k+s}. \tag{5.29}$$

Thus, we improve the loss function (5.28) to encode this sequential information, resulting in

$$\mathcal{L} = \frac{1}{S} \sum_{s=1}^{S} \frac{1}{2(K-s)} \sum_{k=1}^{K-s} \|X_{k+s} - g^s(X_k)\|_2^2 \tag{5.30}$$

for some $S \geq 1$. Notice that the loss function (5.30) aims for g to satisfy $g(X_k) \approx X_{k+1}$, $g^2(X_k) \approx X_{k+2}, \ldots, g^S(X_k) \approx X_{k+S}$, thus providing information to the neural network about S steps of the data into the future. Here the number of steps S becomes another parameter in the neural network that is prescribed by the user prior to training.

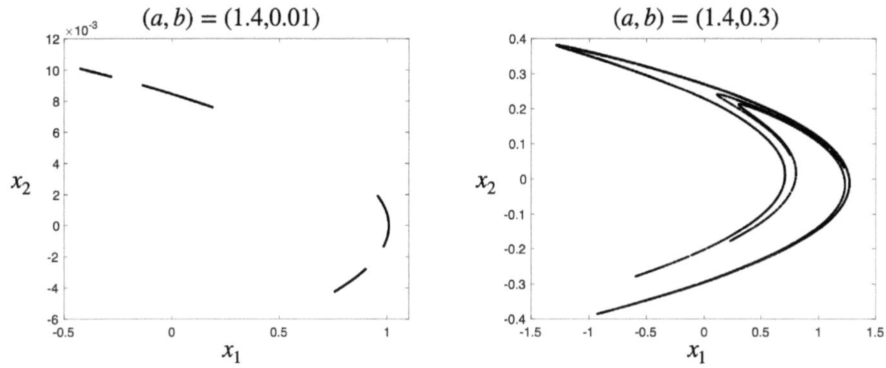

Figure 5.4. *The chaotic attractor of the Hénon map* (5.31) *with* $a = 1.4$ *and* $b = 0.01$ *(left) and* $b = 0.3$ *(right).*

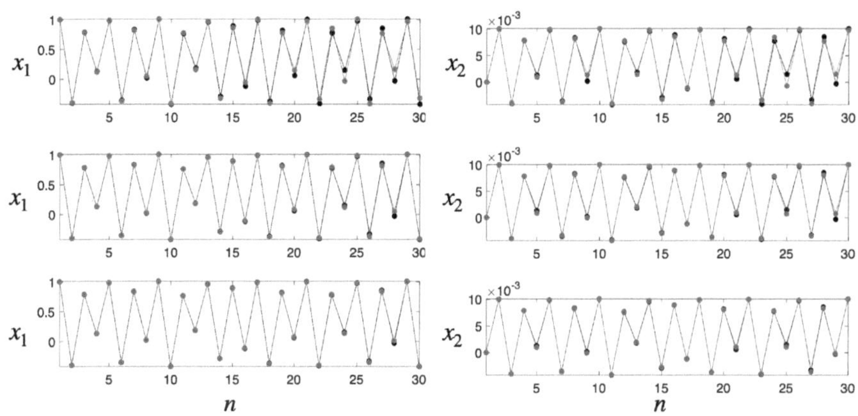

Figure 5.5. *The first* 30 *iterations of the neural network forecast of the Hénon map with* $(a, b) = (1.4, 0.01)$. *On the left are the predictions for* x_1 *(red) against the training data (black), and on the right are* x_2 *(blue). The network is trained by minimizing the loss* (5.30) *with* $S = 1, 2, 3$ *from top to bottom.*

Example 5.3. In this example we will consider data gathered from the Hénon map [118]

$$\begin{bmatrix} x_1 \\ x_2 \end{bmatrix}_{n+1} = \begin{bmatrix} 1 - ax_1^2 + x_2 \\ bx_2 \end{bmatrix}_n, \tag{5.31}$$

parameterized by $(a, b) \in \mathbb{R}^2$. Throughout we fix $a = 1.4$ and consider both $b = 0.01$ and $b = 0.3$. As can be observed in Figure 5.4, $b = 0.01$ leads to a chaotic attractor appearing to be fragments of a single curve in the (x_1, x_2)-plane and displaying little variation in the x_2 direction. Alternatively, $b = 0.3$ is the standard chaotic value for the Hénon map, where the attractor appears to be a single curve folded back and forth on itself in the plane.

We begin with $b = 0.01$. The following demonstrations use a neural network architecture with depth 4 and each layer having a ReLU activation (5.14) and width 20. The training data is gathered from iterating the Hénon map with initial condition $(x_1, x_2) = (1, 0)$ for $K = 1001$ steps into the future. In Figure 5.5 are the predicted orbits of the initial condition $(x_1, x_2) = (1, 0)$ against the training data over 30 iterates of the map. Notice that the network trained with $S = 1$ begins to vary from the truth within 10 iterates, while those with $S = 2, 3$ remain accurate

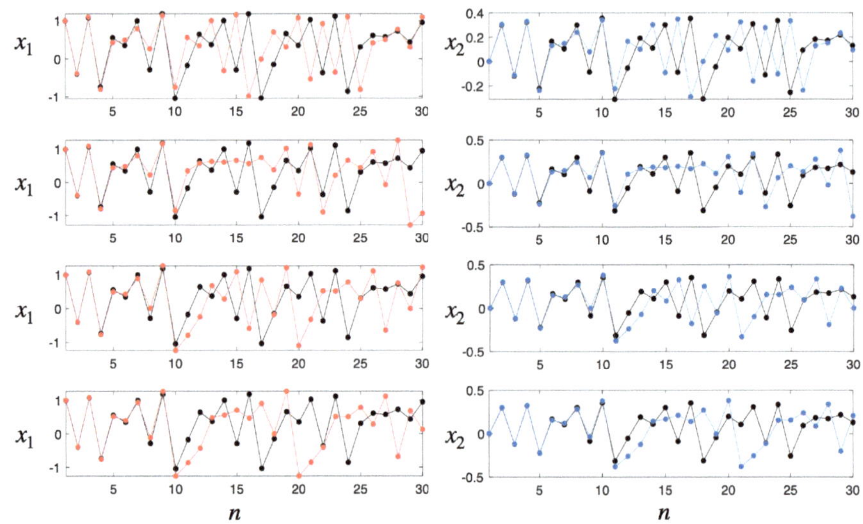

Figure 5.6. *The same as Figure 5.5 except now with parameter values $(a, b) = (1.4, 0.3)$ in the Hénon map. Steps used for the loss function (5.30) are $S = 1, 2, 3, 4$ from top to bottom.*

for almost all 30 iterates. After 30 iterates we have the following errors:

$$\begin{aligned} \underline{S=1}: &\quad \|g^{30}(1,0) - X_{30}\|_2 = \mathtt{0.1060}, \\ \underline{S=2}: &\quad \|g^{30}(1,0) - X_{30}\|_2 = \mathtt{0.0091}, \\ \underline{S=3}: &\quad \|g^{30}(1,0) - X_{30}\|_2 = \mathtt{0.0017}, \end{aligned} \tag{5.32}$$

showing a continual increase in forecasting precision as S increases.

At the parameter value $b = 0.3$ the Hénon mapping dynamics are more complicated and display a higher sensitivity to imprecision by the neural network. This is exemplified in Figure 5.6, where forecast results from the trained neural network using $S = 1, 2, 3, 4$ are provided. One can see a slight improvement by increasing the number of steps S; however, all neural network forecasts become inaccurate within 10 iterations.

One way to measure the quality of the prediction in this case is to compare the number of accurate iterates with the Lyapunov time [26], which is the characteristic timescale in which a system is chaotic. Lyapunov time is defined as the inverse of the largest Lyapunov exponent and formally describes the limit of predictability for a chaotic system. Precisely, it quantifies the number of iterates for nearby initial conditions to separate by a factor of e. For the Hénon map at $(a, b) = (1.4, 0.3)$ the Lyapunov time is between 2 and 3 iterates, meaning that our trained neural network is accurate for at least one Lyapunov time, and in some cases two or three are predicted accurately. Hence, even without being able to forecast the chaotic system far into the future, we have still outperformed the conventional prediction horizon as measured by the Lyapunov time.

Although we can use neural networks to forecast time-series data into the future, they come as a black box. That is, we have no knowledge of the mapping dynamics and therefore cannot interrogate the discovered map beyond simply iterating it. This is certainly a drawback to using neural networks to learn the dynamics of a system—one which is not present when performing symbolic regression onto dictionaries like with SINDy. However, we have already seen that symbolic regression presents its own set of limitations, primarily coming from the choice of library functions and dimension of the data. In Chapter 6 we will see how to combine symbolic regression with neural networks to overcome the limitations presented by both methods.

5.4 ▪ Physics-Informed Neural Networks

We saw in Section 5.2 that the general structure of neural networks allows for fast differentiation of the loss function with respect to the network parameters via backpropagation. This differentiation was employed to perform gradient descent for network training. Following the same procedure, one can also differentiate the neural network with respect to its inputs. This means that the neural network represents a function that is as smooth as the activation functions used to build it. This smoothness of the neural network and the relative ease of differentiation enabled through backpropagation mean that one could also include terms in the loss function that depend on the derivative of the network, thus leading to differential constraints on the network.

The ability to incorporate differential constraints into the loss function of a neural network was identified in early studies by Psichogios and Ungar [230] and Lagaris, Likas, and Fotiadis [164] as a means to use neural networks to approximate solutions to differential equations. Modern advances in computing, the advent of freely available software libraries, and training algorithms such as backpropagation and automatic differentiation [24] have allowed for these ideas to be readapted and updated to reflect the modern computational landscape. Most notable among the recent adaptations of using neural networks to solve differential equations is the work of Raissi, Perdikaris, and Karniadakis [234], which coined the term **Physics-Informed Neural Networks** (PINNs). PINNs have become a highly applicable tool to approximate the solution of ODEs, PDEs, and fractional differential equations [66] while becoming particularly useful in simulating unsteady fluid models at a reduced computational cost [52]. In this section we will primarily follow the original presentation of PINNs [234] to demonstrate the efficacy of approximating solutions of differential equations with neural networks.

We begin by considering the general form of a PDE

$$\begin{aligned} \mathcal{N}[u] &= 0, && x \in \Omega, && t \in [0,T], \\ u(x,t) &= g(x,t), && x \in \partial\Omega, && t \in [0,T], \\ u(x,0) &= f(x), && x \in \Omega, \end{aligned} \tag{5.33}$$

where $\Omega \subset \mathbb{R}^N$ is an open, bounded set with $\partial\Omega$ denoting its boundary and $T > 0$ is the time horizon. The operator $\mathcal{N}$ is a (partial) differential operator, and f and g are given initial and boundary conditions, respectively. Here we are assuming Dirichlet boundary conditions as an example, but the setup is flexible enough to include Neumann or other boundary conditions, as will be demonstrated in Example 5.4 below. The main thrust of a PINN is to approximate the solution of (5.33), $u(x,t)$, as a trained neural network, which we will denote as $\tilde{u}(x,t)$. We emphasize that the inputs are pairs (x,t) and the output is the value of the function $\tilde{u}(x,t)$.

As we have already seen, the key to having a neural network approximate an object is in defining the loss function. The loss function for the PINN approximation $\tilde{u}(x,t)$ is built as the sum of three losses representing (i) the differential equation, (ii) the boundary terms, and (iii) the initial condition. Precisely, we write the loss $\mathcal{L}$ as

$$\mathcal{L} = \mathcal{L}_{\mathrm{u}} + \mathcal{L}_{\mathrm{b}} + \mathcal{L}_0 \tag{5.34}$$

to represent the three components of the sum. First, we consider $J_{\mathrm{u}} \geq 1$ training points $\{(x_j^{\mathrm{u}}, t_j^{\mathrm{u}})\}_{j=1}^{J_{\mathrm{u}}} \subset \Omega \times (0,T]$ to arrive at the differential equation loss:

$$\mathcal{L}_{\mathrm{u}} = \frac{1}{2J_{\mathrm{u}}} \sum_{j=1}^{J_{\mathrm{u}}} \|\mathcal{N}[\tilde{u}(x_j^{\mathrm{u}}, t_j^{\mathrm{u}})]\|_2^2. \tag{5.35}$$

Second, we consider $J_{\mathrm{b}} \geq 1$ points $\{(x_j^{\mathrm{b}}, t_j^{\mathrm{b}})\}_{j=1}^{J_{\mathrm{b}}} \subset \partial\Omega \times [0,T]$ to arrive at the boundary loss:

$$\mathcal{L}_{\mathrm{b}} = \frac{1}{2J_{\mathrm{b}}} \sum_{j=1}^{J_{\mathrm{b}}} \|\tilde{u}(x_j^{\mathrm{b}}, t_j^{\mathrm{b}}) - g(x_j^{\mathrm{b}}, t_j^{\mathrm{b}})\|_2^2. \tag{5.36}$$

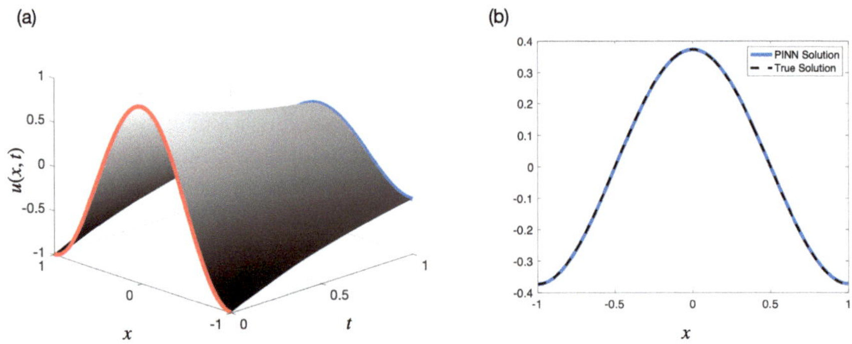

Figure 5.7. (a) *The PINN solution to the heat equation* (5.38). (b) *The PINN solution (blue) compared with the true solution (black, dashed) at* $t = 1$. *The maximal pointwise difference between the two solutions is* $\sim 10^{-3}$.

Finally, the initial condition loss comes from taking $J_0 \geq 1$ points $\{x_j^0\}_{j=1}^{J_0} \subset \Omega$ to get

$$\mathcal{L}_0 = \frac{1}{2J_0} \sum_{j=1}^{J_0} \|\tilde{u}(x_j^0, 0) - f(x_j^0)\|_2^2. \tag{5.37}$$

Notice that all losses are mean square errors. The loss $\mathcal{L}_\mathrm{u}$ enforces that the approximation $\tilde{u}(x,t)$ adheres to the physics of the differential equation, i.e., the governing equations. The losses $\mathcal{L}_\mathrm{b}$ and $\mathcal{L}_0$ enforce that the neural network satisfies the boundary and initial conditions, respectively. We now illustrate with a simple demonstration on the heat equation.

Example 5.4. To demonstrate the utility of PINNs, let us consider the heat equation

$$\begin{aligned} u_t - 0.1u_{xx} &= 0, && x \in (-1,1), \quad t \in [0,1], \\ u_x(\pm 1, t) &= 0, && t \in [0,1], \\ u(x,0) &= \cos(\pi x), && x \in (-1,1). \end{aligned} \tag{5.38}$$

The exact solution to this PDE is given by $u(x,t) = \mathrm{e}^{-0.1\pi^2 t}\cos(\pi x)$, and so we can use this problem as a benchmark for the PINN's performance.

To train the PINN[5] we randomly generate $J_\mathrm{u} = 10000$ points from the interior of the domain, $J_\mathrm{b} = 1000$ points from boundary, and $J_0 = 500$ points at $t = 0$. The network has depth 8, with each layer having width 20 and using a hyperbolic tangent activation function. The loss is as given in (5.34) with the exception that we now have Neumann boundary conditions, so $\mathcal{L}_\mathrm{b}$ becomes

$$\mathcal{L}_\mathrm{b} = \frac{1}{2J_\mathrm{b}} \sum_{k=1}^{J_\mathrm{b}} \|\tilde{u}_x(x_j^\mathrm{b}, t_j^\mathrm{b})\|_2^2. \tag{5.39}$$

The trained PINN solution to (5.38) is presented in Figure 5.7(a) and was trained in 183.3 seconds, resulting in a loss of $\mathtt{4.8671} \times 10^{-5}$. In Figure 5.7 we compare the final state at $t = 1$ of the PINN solution $\tilde{u}(x,t)$ to the exact final state $\mathrm{e}^{-0.1\pi^2}\cos(\pi x)$. The two solutions are nearly indistinguishable to plotting accuracy, as is evidenced by the computation that

$$\sup_{x\in[0,1]} |\tilde{u}(x,1) - \mathrm{e}^{-0.1\pi^2}\cos(\pi x)| = \mathtt{0.0011}. \tag{5.40}$$

[5]The PINN implementations in this example and the one that follows are based on the publicly available code associated to the work [29], which provides a review of neural network methods for solving PDEs.

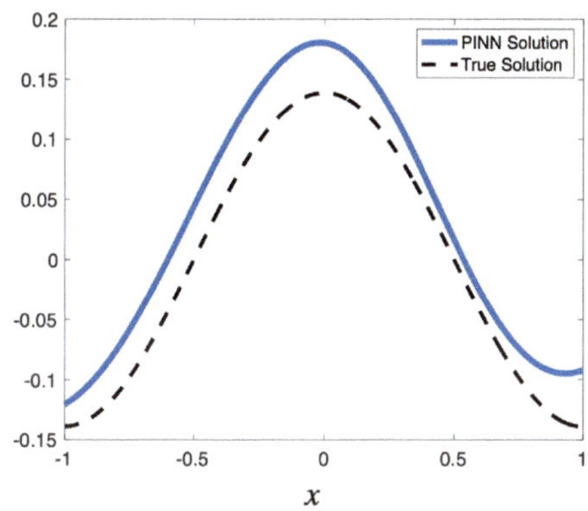

Figure 5.8. *The trained PINN solution (blue) to* (5.38) *compared to the true solution (black, dashed) at* $t = 2$.

Therefore, the PINN provides a fast and accurate approximation of the solution to the diffusion problem (5.38) that can be adjusted to handle a variety of initial and boundary conditions with little more time or computational investment.

One drawback is that we should expect the solutions to be inaccurate beyond the trained time horizon since no knowledge of the solution behavior is passed to the neural network through the loss function. For example, Figure 5.8 shows that evaluating the trained PINN at $t = 2$ results in a solution that varies considerably from the true solution. Notice that the PINN also does not have the appropriate boundary conditions at $t = 2$, again because this information was not passed to the PINN during training.

The next example shows that PINNs can be used for the analysis of traveling waves to reaction-diffusion PDEs. In particular, we use a PINN to solve a singular, parameter-dependent, nonautonomous boundary value problem (BVP). The parameter becomes a trainable variable in the network that is determined through gradient descent, similar to the network weights and biases.

Example 5.5. We consider Nagumo's equation [193], given by the scalar reaction-diffusion PDE

$$u_t = u_{xx} + u(1-u)(u-a), \quad x \in \mathbb{R}, \quad t \geq 0, \tag{5.41}$$

where $u = u(x,t) \in \mathbb{R}$ and $a \in (0,1)$ is a system parameter. The Nagumo equation is known as *bistable* because if we consider the associated nonspatial ODE (by fixing $u_{xx} = 0$), we get

$$u' = u(1-u)(u-a), \tag{5.42}$$

which has exactly three steady states, $u = 0, a, 1$, with both 0 and 1 being stable and a unstable. Thus, (5.42) has two competing stable states with the fate of an initial condition $u(0)$ determined by which side of the unstable state $u = a$ it is on. When space is added into the model, as in the Nagumo equation (5.41), one can find the presence of **traveling wave** solutions to the PDE which asymptotically connect one of the stable states to the other over the spatial medium $x \in \mathbb{R}$.

Precisely, traveling wave solutions to the PDE (5.41) take the form $u(x,t) = U(x - ct)$, with

$$\lim_{\xi \to -\infty} U(\xi) = 0 \quad \text{and} \quad \lim_{\xi \to \infty} U(\xi) = 1. \tag{5.43}$$

Notice that a traveling wave maintains exactly the same shape for all time, given by the profile U, and is simply transported across space at the constant speed of propagation c, as illustrated in

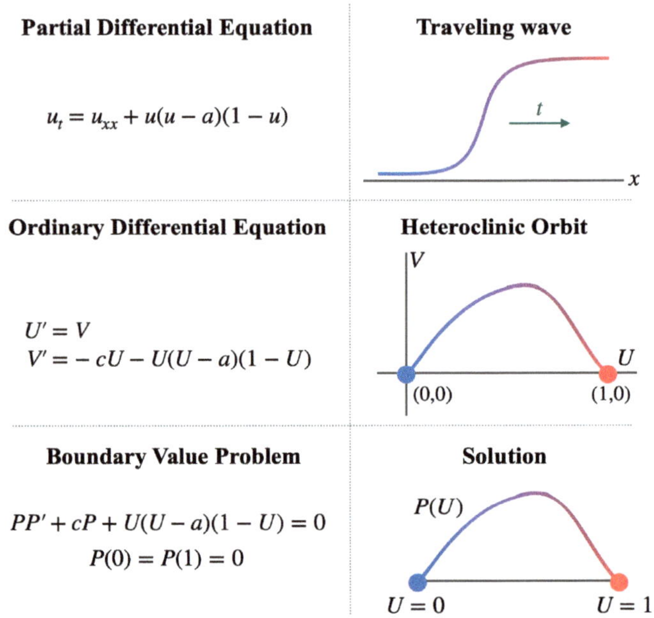

Figure 5.9. *Three equivalent perspectives on traveling wave solutions to Nagumo's equation* (5.41) *formulated using PDEs (top), ODEs (middle), and BVPs (bottom).*

the top panel of Figure 5.9. If $c < 0$, the wave is traveling to the left, while if $c > 0$, it travels rightward. Putting $u(x,t) = U(x-ct)$ into (5.41) results in the second-order ODE for the profile

$$-cU' = U'' + U(1-U)(U-a), \tag{5.44}$$

where $'$ denotes differentiation with respect to the single independent variable $\xi = x - ct$. With (5.44) and the limits (5.43), we transform the search for traveling waves to the Nagumo PDE to finding a heteroclinic orbit in (5.44) connecting $(U, U') = (0,0)$ to $(U, U') = (1,0)$, while c is now a system parameter. This perspective is illustrated in the middle panel of Figure 5.9.

It is proven in [87] that for each $a \in (0,1)$ the following hold:

1. There is a unique value of c for which a heteroclinic orbit connecting $(U, U') = (0,0)$ to $(U, U') = (1,0)$ in (5.44) exists.

2. $U' > 0$ everywhere along the heteroclinic orbit.

3. The unique value of c has the same sign as $\int_0^1 u(1-u)(u-a)du$.

4. When $c > 0$, the PDE (5.41) initiated with any compactly supported initial condition will develop into a pair of diverging traveling waves with asymptotic speed c.

Bistable systems such as (5.41) are often used to model spatially distributed biological populations where small population densities fail to reproduce, leading to extinction ($u = 0$), while large population densities grow to carrying capacity ($u = 1$) [143]. This is called an **Allee effect**. From (4) we see that knowing the value of the wave speed c of the unique traveling wave can predict whether an initially localized population will spread and at what speed.

We now turn to using PINNs to identify the unique wave speed for different a. The search for heteroclinic orbits in (5.44) can be interpreted as a parameter-dependent BVP, albeit with

Table 5.1. *Computed unique traveling wave speeds using a PINN compared with the exact value for various values of $a \in (0,1)$.*

a	PINN wave speed	Exact wave speed
0.1	-0.56719	-0.56569
0.2	-0.42499	-0.42426
0.3	-0.28350	-0.28284
0.4	-0.14152	-0.14142
0.5	-0.00073	0.00000
0.6	0.14285	0.14142
0.7	0.28538	0.28284
0.8	0.42395	0.42426
0.9	0.56773	0.56569

an infinite domain $\xi \in \mathbb{R}$. We will first make the domain compact, which will lead to a BVP on a finite interval that can be solved numerically using a PINN. Since $U' > 0$ for the solution of interest, we can use $U \in [0,1]$ as the independent variable. Following [87], we introduce $P(U) = U'$, bringing (5.44) into the equivalent singular nonautonomous ODE

$$PP' + cP + U(1-U)(U-a) = 0. \tag{5.45}$$

Furthermore, P must satisfy the boundary conditions $P(0) = P(1) = 0$, making (5.45) a BVP with Dirichlet boundary conditions on the finite interval $[0,1]$. This perspective on traveling wave solutions to Nagumo's equation is illustrated at the bottom of Figure 5.9.

This BVP now corresponds to the PINN setting (5.33) with $\mathcal{N}[P]$ taking the form (5.45), $\Omega = (0,1)$, no time dependence, and boundary conditions $P = 0$ at $U = 0, 1$, the boundary of Ω. The training data is comprised of $J_{\rm u} = 1000$ randomly chosen points in $(0,1)$, while $J_{\rm b} = 2$ from the two boundary points $U = 0$ and $U = 1$. The network architecture is taken to have a depth of 12 with each layer having width of 20, all using the scaled exponential linear unit (SELU) activation function,

$$\mathrm{SELU}(x) = \lambda \begin{cases} x, & x > 0, \\ \alpha e^{x} - \alpha, & x \leq 0, \end{cases} \tag{5.46}$$

with [146]

$$\begin{aligned} \lambda &= 1.0507009873554804934193349852946, \\ \alpha &= 1.6732632423543772848170429916717. \end{aligned} \tag{5.47}$$

The SELU function has a number of advantages over other activation functions, primary among them being that the network should be expected to converge faster due to its self-normalizing properties (which will not be covered here) [146, 220, 98, 242]. The wave speed c will be a trainable parameter associated to the neural network whose value is identified during training.

The advantage of working with the Nagumo equation (5.41) is that traveling wave solutions can be found explicitly. Indeed, the traveling waves (U, c) come as sigmoid functions of the form

$$U(\xi) = \frac{1}{2} + \frac{1}{2}\tanh\left(\frac{\xi}{2\sqrt{2}}\right), \qquad c = \frac{1}{\sqrt{2}}(2a-1), \tag{5.48}$$

thus allowing for the comparison of the wave speeds found with the PINN to the exact value for each $a \in (0,1)$. Numerical results are summarized in Table 5.1 for various $a \in (0,1)$. We find agreement between the predicted speed with a PINN and the exact speed to 2–3 significant digits.

5.5 ▪ Discovering Partial Differential Equations

Example 5.5 in the previous section shows that we can include system parameters as trainable pieces of the neural network architecture. The result was that the PINN simultaneously discovered the solution to a nonlinear differential equation while also identifying a coefficient in the equation that gives way to the solution. This concept can be taken a step further to use PINNs for model discovery [233, 234]. The advantage will be that we do not need to approximate spatial or temporal derivatives from data to incorporate them into the library since they can be computed by differentiating the neural network with respect to its inputs. Here we present the general framework of data-driven discovery with PINNs while exploring its advantages and disadvantages.

The setting for model discovery with PINNs is that we are given $J_{\mathrm{d}} \geq 1$ data samples of the solution, $\{u_j^{\mathrm{d}}\}_{j=1}^{J_{\mathrm{d}}} \subset \mathbb{R}^N$, at various points in the space-time domain $\{(x_j^{\mathrm{d}}, t_j^{\mathrm{d}})\}_{j=1}^{J_{\mathrm{d}}} \subset \Omega \times [0,T]$. The goal is to identify a differential operator $\mathcal{N}$ so that a solution u satisfying

$$\mathcal{N}[u] = 0 \quad \forall x \in \Omega,\ t \in [0,T], \tag{5.49}$$

also fits the provided data, $u(x_j^{\mathrm{d}}, t_j^{\mathrm{d}}) = u_j^{\mathrm{d}}$ for all $j = 1, \dots, J_{\mathrm{d}}$. Much like EDMD and SINDy, model discovery with PINNs expands the unknown differential operator $\mathcal{N}$ in a dictionary of candidate terms $\mathcal{D} = \{\psi_1(u), \dots, \psi_K(u)\}$. Notice that our dictionary elements could (and should) include derivatives in space and time, such as u_t, u_{xx}, or uu_x. Then, to simultaneously fit the data u_j^{d} and discover a PDE model for it, we use the loss function

$$\mathcal{L} = \mathcal{L}_{\mathrm{d}} + \mathcal{L}_{\mathrm{u}}. \tag{5.50}$$

The loss term $\mathcal{L}_{\mathrm{d}}$ is used to fit the provided data, given by the mean square error

$$\mathcal{L}_{\mathrm{d}} = \frac{1}{2J_{\mathrm{d}}} \sum_{j=1}^{J_{\mathrm{d}}} \|\tilde{u}(x_j^{\mathrm{d}}, t_j^{\mathrm{d}}) - u_j^{\mathrm{d}}\|_2^2. \tag{5.51}$$

The loss term $\mathcal{L}_{\mathrm{u}}$ is used to discover the physics of the system, given by the mean square error

$$\mathcal{L}_{\mathrm{u}} = \frac{1}{2J_{\mathrm{u}}} \sum_{j=1}^{J_{\mathrm{d}}} \|\xi_1\psi_1(\tilde{u}(x_j^{\mathrm{u}}, t_j^{\mathrm{u}})) + \cdots + \xi_K\psi_K(\tilde{u}(x_j^{\mathrm{u}}, t_j^{\mathrm{u}}))\|_2^2 \tag{5.52}$$

for some $J_{\mathrm{u}} \geq 1$ and domain points $\{(x_j^{\mathrm{u}}, t_j^{\mathrm{u}})\}_{j=1}^{J_{\mathrm{u}}} \subset \Omega \times [0,T]$. The vectors ξ_k are trainable network variables that give way to the PDE model

$$\mathcal{N}[u] = \bar{\xi}_1\psi_1(u) + \cdots + \bar{\xi}_K\psi_K(u) \tag{5.53}$$

for $\bar{\xi}_k$ identified during the training process as the parameter that minimizes the loss $\mathcal{L}_{\mathrm{u}}$.

Despite the theory of model discovery with PINNs being well founded, the inclusion of differential terms in the dictionary means that it can rapidly explode in size. We already saw this issue with EDMD in Section 2.5, but discovering PDEs from data presents another problem as well: nonuniqueness of models. To see this, suppose we are gathering data from the function

$$u(x,t) = \mathrm{e}^{-0.1\pi^2 t}\cos(\pi x). \tag{5.54}$$

We saw in Example 5.4 that this is a solution of the heat equation $u_t = 0.1u_{xx}$. One can readily check that this function also solves the PDEs

$$\begin{aligned} u_t &= -\frac{0.1}{\pi^2}u_{xxxx}, \\ u_t &= -0.1\pi^2 u, \\ u_{tt} &= -0.1\pi^2 u_{xx}, \end{aligned} \tag{5.55}$$

among infinitely many others. All of these models could in theory be discovered using PINNs, depending on what is included in the dictionary. Furthermore, although the above examples are all linear PDEs, we get the same issue with nonlinear PDEs. For example, (5.48) is a solution of the nonlinear Nagumo PDE, while it is also a solution of the wave equation $u_{tt} = c^2 u_{xx}$ and the transport equation $u_t = cu_x$. Notice that overcoming this issue with data from a multiplicity of solutions is not possible in the above PINN discovery framework, as we are using the neural network to approximate only one of the solutions to the PDE, coming from the loss term $\mathcal{L}_{\mathrm{d}}$.

However, for many physical systems we have accurate models derived from the physical laws that govern them. Therefore, knowledge of these governing equations or conservation laws can be leveraged to build the dictionary, meaning that we are using data to determine the coefficients that balance the physical forces. A particular example of this is fluid data, which is governed by the Navier–Stokes equations. PINNs can be used to identify important parameters in the system, such as the Reynolds number [234], or to discover the velocity and pressure fields using only spatiotemporal visualizations of a dye or smoke in the fluid [235]. Beyond just fluids, PINNs have been used to identify system coefficients and parameters for models in solid mechanics [112], power network systems [200], free vibrations [296], and nano-optics and metamaterials [55]. Thus, PINNs find a strong application for performing parameter estimation to fit observations to known governing equations, potentially with incomplete system information.

Example 5.6. We return to the heat equation from Example 5.4. Our training data will be comprised of J_{d} randomly chosen points in the space-time domain $[-1, 1] \times [0, 1]$ from the exact solution $u(x, t) = \mathrm{e}^{-0.1\pi^2 t} \cos(\pi x)$. We will employ a PINN to identify the diffusion coefficient, meaning that we expand the unknown PDE as

$$\mathcal{N}[u] = u_t + \xi_{xx} u_{xx} = 0, \qquad \xi_{xx} \in \mathbb{R}, \tag{5.56}$$

with the exact value of ξ_{xx} being -0.1. Although we have significantly reduced the size of the library, one interpretation of this example is that we are employing a PINN to learn the thermal diffusivity of the medium (ξ_{xx}) using only experimental temperature data (u_j^{d}). Thermal diffusivity of a medium is traditionally measured by the *laser flash method* [218], where the medium is heated from below and a detector on top measures the temperature rise. This toy example presents a potential noninvasive, data-driven alternative to identifying the thermal diffusivity of the medium that makes explicit use of the underlying physics of the system.

In all of the applications that follow we will take $J_{\mathrm{u}} = 10000$ randomly chosen points in $[-1, 1] \times [0, 1]$ for system identification. We summarize the results in the following table:

J_{d}	Loss value	Identified ξ_{xx}	Relative error
50	5.59770×10^{-5}	-0.09625	3.75%
100	4.9529×10^{-5}	-0.10386	3.86%
500	1.92381×10^{-5}	-0.10027	0.27%
1000	8.22741×10^{-6}	-0.10369	3.69%

Even with few training data points we are able to identify the thermal diffusivity with single-digit relative errors. It is also possible to improve upon these values with larger network architectures and/or different parameter initializations. Therefore, these values represent upper bounds on the ability of PINNs to perform parameter estimation, even with relatively little training data.

Now we explore the effect of other terms in the dictionary. We fix $J_{\mathrm{d}} = 500$ for the remaining experiments and start by including the advection term u_x, giving the dictionary expansion

$$\mathcal{N}[u] = u_t + \xi_{xx} u_{xx} + \xi_x u_x = 0, \qquad \xi_x, \xi_{xx} \in \mathbb{R}. \tag{5.57}$$

The trained PINN model has a loss of $\texttt{1.57217} \times 10^{-5}$ and identifies the system coefficients

$$\xi_{xx} = \texttt{-0.10205}, \qquad \xi_x = \texttt{-0.04576}. \tag{5.58}$$

The relative error on ξ_{xx} is 2.05%, while ξ_x is close to the exact value of 0.

Alternatively, we could include the scalar term u in the library, resulting in the dictionary expansion of the differential equation as

$$\mathcal{N}[u] = u_t + \xi_{xx}u_{xx} + \xi_1 u = 0, \qquad \xi_1, \xi_{xx} \in \mathbb{R}. \tag{5.59}$$

One issue here is that the function $u(x,t) = \mathrm{e}^{-0.1\pi^2 t}\cos(\pi x)$ is a solution to the above equation for any ξ_1, ξ_{xx} adhering to the linear relationship

$$-\pi^2\xi_{xx} + \xi_1 = 0.1\pi^2. \tag{5.60}$$

Thus, there are infinitely many possibilities for the system coefficients that the PINN can identify. This degeneracy is exactly what we try to avoid by performing parameter estimation on explicit models derived from known physical laws. Nonetheless, our trained PINN results in a loss of $\texttt{8.08515} \times 10^{-5}$ and identifies the coefficients

$$\xi_{xx} = \texttt{-0.01162}, \qquad \xi_1 = \texttt{0.90052}, \tag{5.61}$$

giving that

$$-\pi^2\xi_{xx} + \xi_1 = \texttt{1.01522}, \tag{5.62}$$

a relative error of 2.86% from the exact value $-0.1\pi^2$. Although we have not identified the heat equation, we have indeed identified an approximate model for the training data provided.

One way to avoid this degeneracy is to incorporate a sparsity condition on the coefficients ξ_{xx} and ξ_1 [35]. This can be done by augmenting the loss function as

$$\mathcal{L} = \mathcal{L}_\mathrm{d} + \mathcal{L}_\mathrm{u} + \lambda(|\xi_{xx}| + |\xi_1|), \tag{5.63}$$

with $\lambda > 0$ another hyperparameter of the system that must be specified before training. A larger value of λ will require smaller $|\xi_{xx}|$ and $|\xi_1|$ to make the loss small, thus putting greater preference on the sparsity condition. Alternatively, a small value of λ preferences fitting the data and the model, while sparsity is not enforced as strongly to make the loss function small. In practice, one should train numerous networks with different values of λ and choose the one that produces the most accurate model, as can be measured by the loss function, prediction accuracy, or any number of other metrics. As an example, taking $\lambda = 0.01$ returns the values

$$\xi_{xx} = \texttt{-0.0836}, \qquad \xi_1 = \texttt{5.19462} \times 10^{-6}. \tag{5.64}$$

We see that the sparsity constraint has done a good job of removing the redundant u term from the PDE, but it has also led to a thermal diffusivity that is smaller in absolute value than the true value of -0.1. This represents a trade-off since our desire for sparsity has enforced that all coefficients are small, not just the one that we knew should not be there.

In the previous example we used synthetic data gathered from the exact solution to the PDE we wished to discover. Of course, this is not a realistic scenario, and the example was only meant to demonstrate some of the advantages and disadvantages of model discovery with PINNs. In practice one has either numerically generated data or real-world data gathered from experiment. Both are expected to result in an inaccurate reading of the true solution to the PDE through either numerical integration error or corrupting noise. Such inaccuracies can present serious problems to the PINN discovery process through the loss component $\mathcal{L}_d$ since the network will attempt to learn the corrupted data, resulting in an overfitting that produces an inaccurate PDE model. Updates to PINNs such as [291, 56] aim to overcome the issue of overfitting noisy data while also being applicable to situations where training data is scarce.

5.6 ▪ More on Neural Networks and PINNs

Our presentation of neural networks in this section was specifically targeted at learning, forecasting, and simulating dynamics. Furthermore, our network layers were all **dense layers**, meaning that the inputs and outputs of each layer are deeply connected through the many nonzero weights. An alternative to dense layers is **convolutional layers**, which apply filters (or kernels) to the input and require fewer parameters than a dense layer. The idea to use filters is similarly motivated by their use in DMD: The input data is extremely high-dimensional and would therefore require an enormous number of parameters to build even a shallow neural network with dense layers. Convolutional neural network layers are commonly used for image data [278], particularly high-resolution images, but have also been used for music recommendations [279], natural language processing [62], and forecasting stock prices [274]. Neural networks with convolutional layers are also a natural fit for the high-dimensional discrete space-time data and so are being employed for discovering, solving, and forecasting PDEs [241, 266, 294, 207, 260, 298].

Since 2019, the literature on PINNs has exploded. This is evinced by reviews written in 2021 [29] and 2022 [66], along with another review in 2022 exclusively focused on the application of PINNs to fluid mechanics [52]. These reviews point to the numerous successes of applying PINNs to both simulate and discover nonlinear PDEs. Notable extensions of the method include PINNs for stiff systems [131], nonlocal models [214], and the discovery of parameterized PDEs [221]. Furthermore, extended PINNs (XPINNs) are a generalization of PINNs that involves multiple neural networks to decompose the domain for parallelization in space and time [130, 128]. Thus, the power and applicability of PINNs only continues to grow.

Rigorous convergence results are also beginning to follow the many experiments with PINNs. For linear second-order elliptic and parabolic PDEs, the work [254] proves that as the amount of data used to train the PINN $(J_{\mathrm{u}}, J_{\mathrm{b}}, J_0)$ grows without bound, the PINN solution strongly converges to the true PDE solution. Similarly, [132] establishes upper bounds on the number of training samples, depth, and width of the PINN to achieve a desired accuracy for solutions to linear second-order elliptic PDEs. Finally, [199] provides rigorous bounds on the generalization error in terms of the training error and the number of training samples used.

There are also new approximation results that assert the existence of deep neural network architectures and training procedures to arbitrarily approximate functions [5, 2]. Results of this nature are termed *practical existence theorems* and are inspired by similar approximation results in compressive sensing [3, 4]. They offer explicit network architectures for function approximation and explicit error bounds in terms of the amount of data for network training. The proofs of these practical existence theorems rely on established results that neural networks can approximate polynomials to arbitrary accuracy [79]. Thus, fixing a polynomial degree and using individual neural networks to approximate each monomial up to the given degree within a desired error tolerance leaves one only to tune the coefficients in the span of the monomials that give the network output. These results may soon be mature enough to provide convergence results on PINN architectures that guarantee their performance.

Chapter 6

Autoencoder Neural Networks

In this chapter we employ neural networks to provide optimal changes of variable to recast dynamics in simpler forms. The idea is to mimic topological conjugacies whose application in dynamical systems goes back almost to the beginning. Even without the proper mathematical underpinning, the idea of recasting dynamics in a simpler frame of reference has existed for centuries. For example, the geocentric model of the solar system was espoused by great thinkers such as Aristotle and Ptolemy and puts Earth at the center with the Sun and planets moving around it. As shown in Figure 6.1, the incommensurate periods of rotation of each planet about the Sun leads to epicyclic motion when observed from Earth. However, moving to a coordinate frame with the Sun at the center—a heliocentric frame—leads to all of the motions of the planets lying along ellipses. Moving between geocentric and heliocentric frames provides a tangible example of a simplifying coordinate transformation in which all motion is simple and predictable.

The goal of this chapter is to use neural networks to match systems or data to simpler models. In many applications these transformations exist only in theory, and so not having access to their explicit definition means that we cannot use them beyond the theoretical realm. Thus, the work in this chapter can be used to enhance the existing mathematical theory of dynamical systems by learning these transformations. Moreover, the intuition that is gained by studying existing models can be carried over to interpreting data provided without an underlying model. Here we will see how neural networks can complement and extend well-developed mathematical techniques and bridge the gap between prototypical models and phenomena observed in experiment.

We begin this chapter with an exposition of topological conjugacies which are the mathematical framework that underpins all of the computational methods in this section. We then proceed to demonstrate how the autoencoder neural network architecture can be used to approximate conjugacies between systems. From there we move on to a number of experiments that couple optimal variable transformations using neural networks with simplified models. The identified simplified models will be problem dependent, and we focus on diverse applications, such as normal forms, global linearizations, and coupling autoencoders with model discovery.

6.1 ▪ Topological Conjugacies

In this section we outline the theory that provides the mathematical basis for the computational methods studied in this chapter. We begin with two continuous functions $f : \mathcal{X} \to \mathcal{X}$ and $g : \mathcal{Y} \to \mathcal{Y}$ mapping topological spaces $\mathcal{X}$ and $\mathcal{Y}$, respectively, back into themselves. The functions are said to be **topologically conjugate**, or just conjugate, if there exists a continuous bijection with a continuous inverse (a homeomorphism) $h : \mathcal{X} \to \mathcal{Y}$ so that $h \circ f = g \circ h$. The

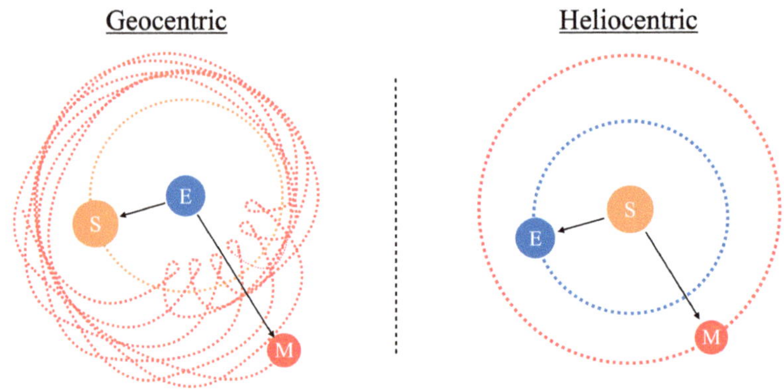

Figure 6.1. *The motion of the Sun (yellow), the Earth (blue), and Mars (red) in a geocentric (left) versus heliocentric (right) coordinate frame.*

function h is the **conjugacy** between functions f and g and equivalently provides that

$$f = h^{-1} \circ g \circ h. \tag{6.1}$$

The relationship (6.1) allows one to arrive at $f(x)$ for any $x \in \mathcal{X}$ by first transforming x to a unique point $h(x) \in \mathcal{Y}$, then applying g to get a new point $g(h(y)) \in \mathcal{Y}$, and then transforming back to $\mathcal{X}$, giving exactly $h^{-1}(g(h(x))) = f(x)$. The conjugacy diagram

$$\begin{array}{ccc} \mathcal{X} & \xrightarrow{f} & \mathcal{X} \\ h \updownarrow & & \updownarrow h \\ \mathcal{Y} & \xrightarrow{g} & \mathcal{Y} \end{array} \tag{6.2}$$

provides a pictorial representation of the identity (6.1) between conjugate functions f and g.

Conjugacies define an equivalence relation on the space of all continuous surjections from a topological space back to itself. This means that we can organize functions into equivalence classes, all of which essentially act in exactly the same way on their respective domains. Importantly, the domains $\mathcal{X}$ and $\mathcal{Y}$ need not be the same space, only that the function h transforms the space into the other while preserving its topological organization. This comes from the fact that h is a homeomorphism from $\mathcal{X}$ to $\mathcal{Y}$. Furthermore, the relationship (6.1) means that the reorganization of $\mathcal{X}$ from the application of f and of $\mathcal{Y}$ from the application of g is again preserved through h and its inverse. This is important to note since h is telling us not only something about how to move between the spaces $\mathcal{X}$ and $\mathcal{Y}$ but also how to move between the functions f and g.

The equivalence relation defined by topologically conjugate functions plays an important role in the study of dynamical systems, as it can be used to gain insight into a potentially complex system from one that is simpler. Precisely, let us suppose $f : \mathcal{X} \to \mathcal{X}$ and $g : \mathcal{Y} \to \mathcal{Y}$ are used to define discrete dynamical systems

$$x_{n+1} = f(x_n) \tag{6.3}$$

and

$$y_{n+1} = g(y_n) \tag{6.4}$$

for all $n \geq 0$ and further assume there exists a conjugacy $h : \mathcal{X} \to \mathcal{Y}$ between f and g. Given an arbitrary initial condition $x_0 \in \mathcal{X}$, its orbit under f is the infinite sequence $\{x_0, x_1, x_2, \dots\}$. Then applying h to each element in the sequence gives $\{h(x_0), h(x_1), h(x_2), \dots\}$, a sequence in $\mathcal{Y}$. Furthermore, using the conjugacy relationship between f and g we find that $h(x_{n+1}) = h(f(x_n)) = g(h(x_n))$ for all $n \geq 0$. Not only does the sequence $\{h(x_0), h(x_1), h(x_2), \dots\}$ belong to $\mathcal{Y}$, but it is *exactly* the orbit of $y_0 := h(x_0)$ under g. This means that orbit sequences of f and g are mapped into each other in a one-to-one manner by the conjugacy h. Precisely, if x_0 is a fixed point of f, meaning $x_n = x_0$ for all $n \geq 1$, then $h(x_0)$ is a fixed point of g. Similarly, if the sequence $\{x_0, x_1, x_2, \dots\}$ is a p-cycle, meaning $x_p = x_0$ for some $p \geq 1$, then $\{h(x_0), h(x_1), h(x_2), \dots\}$ is a p-cycle of g. This correspondence continues by relating any and all dynamical features of f to g and back. In summary, the dynamics of f and g are identical, even if they do not have the same domains and/or look very different.

Example 6.1. A famous example of topologically conjugate dynamical systems is the tent map

$$x_{n+1} = 2\min\{x_n, 1 - x_n\} \tag{6.5}$$

and the logistic map

$$y_{n+1} = 4y_n(1 - y_n). \tag{6.6}$$

Both mappings have domain $\mathcal{X} = \mathcal{Y} = [0, 1]$, and both are chaotic according to Devaney's definition [73]. This means they have infinitely many periodic points that densely fill their domains, exhibit sensitive dependence on initial conditions, and have an orbit that will eventually visit any open subset of their domain. In fact, one way to prove this is by identifying all of these characteristics for the simpler tent map (6.5) and then establishing a conjugacy which will transfer these properties over to the logistic map (6.6).

Prior to describing the topological conjugacy, let us explore the implications of having a conjugacy between these mappings. Although the logistic map is a relatively simple quadratic map, it can still be difficult analyze. For example, identifying the period p points of the logistic map requires finding roots of degree 2^p polynomial, coming from the search for fixed points to the pth composition of the logistic map with itself. Alternatively, one can check with relative ease that all periodic points of the tent map are rational numbers in $[0, 1]$ taking the form

$$x = \frac{2q}{2^p + s}, \tag{6.7}$$

where $s = \pm 1$ and $0 \leq 2q \leq 2^p + s$. An advantage of having a conjugacy between the tent and logistic maps is that one can transform the known period p points of the tent map to obtain each and every period p point of the logistic map, even for very large values of p. Beyond just periodic points, one can transfer other properties between conjugate systems, including topological entropy, symbolic dynamics, transitive orbits, and whether the system is chaotic [73].

Now we return to identifying the conjugacy between (6.6) and (6.5). The existence of a conjugacy h requires that

$$h(2x) = 4h(x)(1 - h(x)), \qquad x \in [0, 1/2], \tag{6.8}$$

and

$$h(2 - 2x) = 4h(x)(1 - h(x)), \qquad x \in (1/2, 1]. \tag{6.9}$$

Equation (6.8) is satisfied by functions $h(x) = \sin^2(\alpha x)$, coming from the double-angle formula

$$\sin^2(2\alpha x) = 2\sin^2(\alpha x)\cos^2(\alpha x) \tag{6.10}$$

and the Pythagorean identity

$$4\sin^2(\alpha x)(1 - \sin^2(\alpha x)) = 4\sin^2(\alpha x)\cos^2(\alpha x). \tag{6.11}$$

We further require h to map the interval $[0,1]$ back into itself in a one-to-one manner, and so we require $\alpha = \pi/2$. Finally, the choice $h(x) = \sin^2(\pi x/2)$ satisfies (6.9) since

$$h(2-2x) = \sin^2(\pi - \pi x) = \sin^2(\pi x) = h(2x), \tag{6.12}$$

which now replicates the previous analysis for $x \in [0, 1/2]$. Thus, we have identified a conjugacy between the tent and logistic maps, given by

$$h(x) = \sin^2\left(\frac{\pi x}{2}\right), \qquad x \in [0,1]. \tag{6.13}$$

A simple application of the conjugacy (6.13) is to identify fixed points of the logistic map using those of the tent map. The tent map has fixed points $x = 0, 2/3$, whose images under the conjugacy (6.13) are $h(0) = 0$ and $h(2/3) = 3/4$, which one can readily verify are fixed points of the logistic map. Furthermore, solving $y = 4y(1-y)$ gives $y = 0, 3/4$ as the only two fixed points of the logistic map, showing that all fixed points are found through the tent map.

What the tent and logistic maps have in common is a single extremum. Such maps are referred to as unimodal, and Feigenbaum famously documented that parameter-dependent unimodal maps appear to have the same qualitative behavior [86], regardless of how simple or complicated they appear. This was made precise by the work of Milnor and Thurston, who demonstrated that under mild assumptions every unimodal map is topologically conjugate to the logistic map

$$y_{n+1} = r y_n(1-y_n) \tag{6.14}$$

for some value of $r \in [0,4]$ [197]. Above we showed this for the tent and logistic maps with $r = 4$. The importance here is that the logistic map (6.14) is one of the most well-studied models in all of dynamical systems, and therefore identifying a conjugacy between it and another map of interest allows one to transfer the knowledge of the logistic map to the conjugate mapping.

In Example 6.1 we were introduced to a *global* conjugacy since the entire state space of one map is transformed to the entire state space of another. Beyond a few isolated examples, global conjugacies are often difficult to identify, and even abstractly proving the existence of a global conjugacy is a difficult task. One can alternatively search for a *local* conjugacy that simply identifies a conjugacy between functions after restricting their domains to a small region of state space. In this case conjugacies are often proven to exist by restricting the domains of the functions to a small enough region of space and applying fixed point theorems to prove that local conjugacies exist without ever knowing what they look like. Examples of local conjugacies abound in dynamical systems theory and include local linearization results and normal forms [109]. The following example provides one such local linearization result, the Hartman–Grobman theorem.

Example 6.2. Consider a smooth mapping

$$x_{n+1} = f(x_n), \tag{6.15}$$

with $f : \mathbb{R}^d \to \mathbb{R}^d$, and suppose that x_* is a fixed point, i.e., $x_* = f(x_*)$. Define $A := Df(x_*) \in \mathbb{R}^{d\times d}$ to be the Jacobian matrix of f evaluated at the fixed point x_*. Then the Hartman–Grobman theorem [109, 232] states that if A has no eigenvalues on the unit circle in the complex plane (it is **hyperbolic**), there is an open neighborhood $U \subset \mathbb{R}^d$ of x_* so that $f : U \to \mathbb{R}^d$ is topologically conjugate to the linear map

$$y_{n+1} = Ay_n. \tag{6.16}$$

Thus, the Hartman–Grobman theorem provides an example of a local conjugacy, as the equivalence between the general nonlinear map (6.15) and the linear map (6.16) can only be shown to exist when f is restricted to a small neighborhood of the fixed point x_*.

There is some intuition that explains why the Hartman–Grobman theorem can only hold as a local conjugacy. Since A has no eigenvalues on the unit circle of the complex plane, (6.16) has exactly one fixed point given by $y = 0$. Since conjugacies map fixed points into each other, this fixed point must correspond to the fixed point $x_* \in U$ in the nonlinear system (6.15). If f had another fixed point $x_{**} \neq x_*$ in U, then the conjugacy would require it to be mapped to a fixed point of (6.16). But, since $y = 0$ is the only fixed point of (6.16), this means that x_{**} and x_* are mapped to the same point, violating that the conjugacy is a bijection. Thus, f can only have one fixed point in U, and the size of U is limited by the distance of x_* to other fixed points of (6.15).

Let us now briefly outline the proof of the Hartman–Grobman theorem, as it demonstrates how conjugacies can be shown to exist without identifying them explicitly. Without loss of generality we assume $x_* = 0$ by simply shifting all x by x_* to bring the fixed point to the origin. We write

$$f(x) = Ax + F(x), \tag{6.17}$$

where $F(x) = f(x) - Ax$, satisfying $F(0) = 0$ and $DF(0) = 0$. Write the desired conjugacy as

$$h(x) = x + H(x), \tag{6.18}$$

with H continuous and satisfying $H(0) = 0$, constituting a near-identity transformation when $\|x\|$ is small. Since A has no eigenvalues on the unit circle, we can write $\mathbb{R}^d = E^s \oplus E^u$, where E^s is the generalized eigenspace of A for eigenvalues whose moduli are less than 1 and E^u is the generalized eigenspace for eigenvalues with moduli greater than 1. The result is that any $x \in \mathbb{R}^d$ can be written $x = x_s + x_u$, where $x_s \in E^s$ and $x_u \in E^u$. We similarly split H into $H_s : E^s \to E^s$ and $H_u : E^u \to E^u$ so that our conjugacy is now $h(x) = x + H_s(x_s) + H_u(x_u)$.

The goal is to have the conjugacy h satisfy $h \circ (A + F) = A \circ h$. By definition of a conjugacy, we require one of the following equivalent conditions to be true:

$$\begin{aligned} H &= -F \circ (A + F)^{-1} + A \circ H \circ (A + F)^{-1}, \\ H &= A^{-1} \circ F + A^{-1} \circ H \circ (A + F). \end{aligned} \tag{6.19}$$

Both conditions define a fixed point problem for the nonlinear component of the conjugacy, H. The first equation is enforced for all $x_s \in E^s$ to obtain H_s, and the second is enforced for all $x_u \in E^u$ to obtain H_u. The existence of H_s and H_u as fixed points to these equations comes from an application of the contraction mapping principle, for which we are required to keep $\|x_s\|$ and $\|x_u\|$ small to guarantee the map is indeed contracting. Finally, the fact that $h(x) = x + H(x)$ is a homeomorphism can be shown to follow from $H(0) = 0$ and the continuity of H but may require further restricting the neighborhood U to be even smaller.

In some cases, once a local conjugacy is established, its region of validity can be identified and extended. This is the case for the Hartman–Grobman theorem of Example 6.2, where [165] extends the linearization to the entire basin of attraction when all eigenvalues of the matrix A have modulus smaller than 1. Importantly, both global and local conjugacies can be shown to exist without ever knowing what the transformation between the systems actual is. This means that the conjugacy between systems can be exploited in theory but not in practice. In the sections that follow we demonstrate how neural networks can be used to identify these conjugacies, lifting ourselves from the theoretical realm to a more practical one.

6.2 ▪ Autoencoders as Approximate Conjugacies

As we saw in the previous section, conjugacies can provide simplifying coordinate transformations that preserve the dynamical structure of a system. They can be used to understand and classify chaotic dynamics or to show that iterates in the neighborhood of a hyperbolic fixed point are equivalent to the linearized dynamics at the fixed point. Furthermore, we saw that the existence of local conjugacies can sometimes be established through fixed point iteration in function spaces, thus giving the existence of a conjugacy without knowing exactly what it is. In this section we will demonstrate how conjugacies can be identified using neural networks.

Originally conceived as a method of unsupervised nonlinear dimensionality reduction [155], **autoencoder neural networks** are a neural network architecture that can be used to approximate topological conjugacies from data. As we saw in Chapter 5, a neural network is just a function whose output is prescribed by a loss function. In this way, an autoencoder is fully described by its loss function, which in this case is meant to have it approximate a conjugacy between functions.

Start by supposing we have functions $f : \mathcal{X} \to \mathcal{X}$ and $g : \mathcal{Y} \to \mathcal{Y}$ which are known to be conjugate. The conjugacy $h : \mathcal{X} \to \mathcal{Y}$ in this case may be unknown, but since h is a function we can approximate it by a neural network, denoted by η. However, h is required to be a homeomorphism, and although building in invertibility to a neural network can be done [129, 17], it is difficult. We relax the condition of invertibility of η by simultaneously learning an approximation of the inverse of h with a neural network, denoted ρ. With data points $\{x_k\}_{k=1}^K \subset \mathcal{X}$ gathered to train the neural network, we enforce that ρ is the inverse of η through the loss

$$\mathcal{L}_1 = \frac{1}{2K}\sum_{k=1}^{K} \|\rho(\eta(x_k)) - x_k\|_2^2. \tag{6.20}$$

The loss $\mathcal{L}_1$ enforces that applying η and then ρ to each data point x_k returns x_k again, so $\rho \circ \eta$ is approximately the identity function. Without accounting for any dynamics of the problem, $\mathcal{L}_1$ constitutes the standard loss for an unsupervised autoencoder whereby one searches for a coordinate transformation η, termed the **encoder**, to a space of features which is often of a lower dimension. The function ρ, termed the **decoder**, undoes the coordinate transformation and lifts one back from the coded space to the original observation space of the data $\{x_k\}_{k=1}^K$.

Although (6.20) provides the standard loss for unsupervised learning with an autoencoder, it does not capture any of the dynamics of f and g, and so η is not yet an approximation of a conjugacy between them. Recalling that a conjugacy h satisfies $f = h^{-1} \circ g \circ h$, with the network approximations η and ρ of h and h^{-1}, this becomes $f = \rho \circ g \circ \eta$. Moreover, composing f with itself $s \geq 1$ should further give $f^s = \rho \circ g^s \circ \eta$. Thus, for a fixed $S \geq 1$ we introduce a second loss function that captures the conjugacy relationship between f and g:

$$\mathcal{L}_2 = \frac{1}{2SK}\sum_{s=1}^{S}\sum_{k=1}^{K} \|\rho(g^s(\eta(x_k))) - f^s(x_k)\|_2^2. \tag{6.21}$$

The loss $\mathcal{L}_2$ enforces the conjugacy relationship between f and g while also encoding the forward iterative dynamics up to some fixed number of steps S. This loss captures moving from the top left to the top right in the commutative diagram (6.2).

Finally, we introduce a third loss function that captures the dynamics in the image of η. This is represented by iterating the function g and pictorially moving from the bottom left to the bottom right of the commutative diagram (6.2). Again with a fixed $S \geq 1$ we have the loss

$$\mathcal{L}_3 = \frac{1}{2SK}\sum_{s=1}^{S}\sum_{k=1}^{K} \|g^s(\eta(x_k)) - \eta(f^s(x_k))\|_2^2, \tag{6.22}$$

which is used to enforce the topological organization of the state space of g. The total loss function we enforce for the autoencoder to approximate a topological conjugacy is

$$\mathcal{L}_{\text{auto}} = \mathcal{L}_1 + \mathcal{L}_2 + \mathcal{L}_3. \tag{6.23}$$

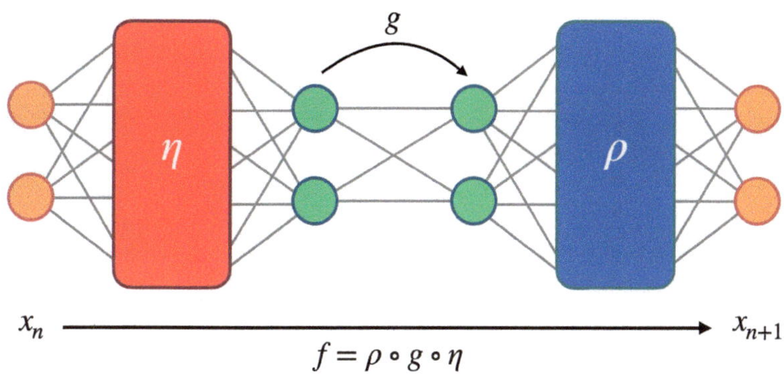

Figure 6.2. *An illustration of an autoencoder neural network approximating a conjugacy between dynamical systems $x_{n+1} = f(x_n)$ and $y_{n+1} = g(y_n)$. The training process learns an encoder, η, and a decoder, ρ, that approximate the conjugacy function and its inverse, respectively, so that the mapping f is approximated by $\rho \circ g \circ \eta$.*

With the approximate conjugacy functions η and ρ we replace the true commutative diagram (6.2) with the pair of autoencoder commutative diagrams

$$\begin{array}{ccc} \mathcal{X} & \xrightarrow{f} & \mathcal{X} \\ \eta\downarrow & & \downarrow\eta \\ \mathcal{Y} & \xrightarrow{g} & \mathcal{Y} \end{array} \qquad \begin{array}{ccc} \mathcal{X} & \xrightarrow{f} & \mathcal{X} \\ \uparrow\rho & & \rho\uparrow \\ \mathcal{Y} & \xrightarrow{g} & \mathcal{Y} \end{array} \tag{6.24}$$

representing the unidirectional nature of the encoder η and the decoder ρ. The network is illustrated in Figure 6.2.

Example 6.3. Let us return to the tent and logistic maps of Example 6.1 from the previous section. We saw that the conjugacy between these two maps is known and given by

$$h(x) = \sin^2\left(\frac{\pi x}{2}\right) \tag{6.25}$$

for all $x \in [0, 1]$. Since this conjugacy is known, we will use it to gauge the ability of autoencoder neural networks to discover conjugacy functions and their inverses.

The neural network is trained by minimizing the loss function (6.23) using $S = 1$ and $K = 1000$ randomly selected points from the interval $[0, 1]$. The encoder and decoder are both composed of six layers, each with a width of 200. Figure 6.3 plots both the discovered encoder and decoder functions against the exact conjugacy and its inverse, respectively. The maximal pointwise absolute difference between the encoder and the true conjugacy over the 1000 linearly spaced points between 0 and 1 used to produce the figure is `0.00855`, while the absolute difference for the decoder and the true inverse of the conjugacy is `0.03271`, showing strong performance of the autoencoder to approximate topological conjugacies.

Example 6.4. The tent map (6.5) is conjugate to many other unimodal mappings beyond just the logistic map. However, outside of a handful of simple examples, like the logistic map, the conjugacy is not known explicitly. This is the case for the sine map

$$y_{n+1} = \sin(\pi y_n), \tag{6.26}$$

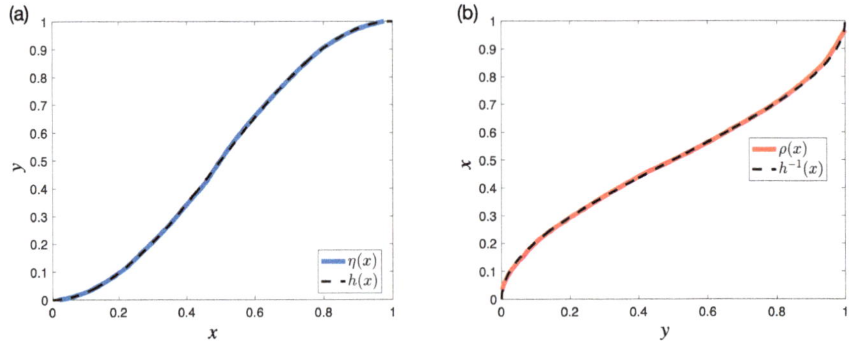

Figure 6.3. *Results of a trained autoencoder to find a conjugacy between the tent and logistic maps.* (a) *The encoder function* $\eta(x)$ *(blue) approximates the true conjugacy* $h(x) = \sin^2(\pi x/2)$ *(black).* (b) *The decoder function* $\rho(x)$ *(red) approximates the true inverse of the conjugacy* $h^{-1}(y) = 2\arcsin(\sqrt{y})/\pi$.

Table 6.1. *Using the trained autoencoder we can identify the location of periodic points in the sine map using those in the tent map as given in* (6.7). *The tent map periodic points are given in the* x *column, while each corresponding periodic point in the sine map is given by its image under the encoder* η *and compared with the true value* $h(x)$. *Here* h *is meant to denote the image of each* x *under the exact conjugacy known to exist between the tent and sine mappings.*

Period	x	$\eta(x)$	$h(x)$
1	0	0.00814	0.00000
1	2/3	−0.73994	−0.73648
2	2/5	−0.34760	−0.34729
2	4/5	−0.88354	−0.88710
3	2/9	−0.13599	−0.13335
3	2/7	−0.20847	−0.21209
3	4/9	−0.40774	−0.40679
3	4/7	−0.61166	−0.61809
3	6/7	−0.93321	−0.93197
3	8/9	−0.95547	−0.95743

where the domain can be taken to be $[0, 1]$ or $[-1, 0]$ since the odd symmetry of the map guarantees that it generates identical dynamics on each interval. In this case we can again use an autoencoder neural network to identify the topological conjugacy that is known to exist between the tent and sine maps. As an aside, we note that since topological conjugacies form an equivalence relation, it follows by transitivity that the logistic and sine maps are also conjugate. Therefore, an alternative exercise would be to employ an autoencoder to identify the conjugacy between the logistic and sine maps or simply use the composition of encoders and decoders found in this example and the previous one to produce this (numerical) conjugacy.

All network specifications are identical to the previous example. We do not impose the domain of the sine map to be either of $[-1, 0]$ or $[0, 1]$ but provide the neural network freedom to identify either for the conjugacy. The trained network for this example identifies $[-1, 0]$, with plots of the encoder and decoder provided in the accompanying notebooks for this section.

We check the validity of this numerically obtained conjugacy with the location of the periodic points in the tent and sine maps. We present our results in Table 6.1 where we identify all periodic points of the tent map, denoted x, up to period 3 and provide their corresponding image under the encoder η. We further provide the value of the corresponding periodic points of the sine map from numerical solutions of $S^p(y) = y$ with $p = 1, 2, 3$ and $S(y)$ denoting the sine map (6.26). The column of periodic points of S are headed by $h(x)$ since they are the image of the periodic points of the tent map under the topological conjugacy h. In all but one case results agree to two significant digits, demonstrating the efficacy of the conjugacy for identifying periodic points in the conjugate map. This strength will be used in more detail in Section 6.5 to gain information about the topological organization of a chaotic attractor in a continuous-time dynamical system. Another check that has been omitted for brevity would be to take the values $h(x)$ in Table 6.1 and apply the decoder function ρ to them to approximate the periodic points of the tent map.

As a final remark, the presence of a period 3 orbit for mappings from an interval back to itself is a classic criterion for chaos [174], demonstrating that the sine map is chaotic.

6.3 ▪ Normal Forms and Bifurcations

The theme of this chapter is using autoencoders to provide nonlinear changes of variables that bring systems into more interpretable forms. Often these interpretable representations of dynamics come as **normal forms**. Familiar examples from the study of dynamical systems include the Jordan normal form, which block diagonalizes matrices according to their spectrum, and the Hartman–Grobman theorem (see Example 6.2), which demonstrates the local equivalence of a nonlinear system near a hyperbolic fixed point to a (canonical) linear one. Normal forms typically represent well-studied systems that guide the understanding of more complex ones, even in cases where a direct equivalence through topological conjugacy cannot necessarily be made.

In this section we focus on bifurcations and some of the well-known topological conjugacies that exist between systems exhibiting them. Much like the Hartman–Grobman theorem, the topological conjugacies we consider here will be local, as they can only hold in a neighborhood of a bifurcation. Moreover, the normal forms we consider help to fill in the gaps of the Hartman–Grobman theorem, as it only applies to hyperbolic equilibria. The bifurcations we study arise from changes in stability of a fixed point as one introduces small changes to a system parameter, with the bifurcation occurring exactly at the point where the fixed point is not hyperbolic, therefore covering where Hartman–Grobman is inapplicable.

The general method of this section requires the consideration of parameter-dependent diffeomorphism

$$x_{n+1} = f(x_n, \mu), \tag{6.27}$$

with a sufficiently smooth function $f : \mathbb{R}^d \times \mathbb{R}^p \to \mathbb{R}^d$. The goal will be to use autoencoders to identify a coordinate transformation into a normal form

$$y_{n+1} = g(y_n, \nu), \tag{6.28}$$

which can be used to locally analyze the bifurcation structure of f. The locality of the transformation will be in reference to an assumed fixed point $(x, \mu) = (x^*, \mu^*)$ of f, for which the linearization with respect to x at this point, denoted $f_x(x^*, \mu^*)$, has exactly one eigenvalue at ± 1 and all other eigenvalues off of the unit circle in the complex plane. By not considering any other eigenvalues on the unit circle we are considering the simplest cases of bifurcations that can take place. This is primarily due to the fact that these bifurcations are the most well studied and are accompanied by rigorous results of topological conjugacy to link them with specific normal forms [206, 109, 162, 97]. Furthermore, all of these results extend to systems in continuous time as well, with the unit circle being replaced by the imaginary axis for the nonhyperbolicity condition. Let us illustrate with a simple example.

Example 6.5. We illustrate by example by considering the one-dimensional map

$$x_{n+1} = (1+\mu)x_n - \mathrm{e}^{x_n}. \tag{6.29}$$

A fixed point of this map is a solution of the transcendental equation

$$\mu x - \mathrm{e}^x = 0, \tag{6.30}$$

while a bifurcation takes place when the linearization has an eigenvalue of 1, meaning

$$\mu - \mathrm{e}^x = 0. \tag{6.31}$$

Combining these two conditions gives $(x^*, \mu^*) = (1, \mathrm{e})$ as the location of a nonhyperbolic equilibrium of (6.29). Replacing e^x with a Taylor expansion about $x = 1$, we approximate the two branches of fixed points emerging from the bifurcation point into $\mu > \mathrm{e}$ by

$$x^*_\pm(\mu) \approx 1 \pm \sqrt{\frac{2}{\mathrm{e}}(\mu - \mathrm{e})}. \tag{6.32}$$

In a neighborhood to the left of $\mu = \mathrm{e}$ there are no equilibria of (6.29), while slightly to the right there there are two. Furthermore, $x_+(\mu)$ is stable, and $x^*_-(\mu)$ is unstable, at least in a neighborhood to the right of $\mu = \mathrm{e}$. The result is that a **saddle-node bifurcation** takes place in (6.29) wherein two fixed points spontaneously emerge at $(x, \mu) = (1, \mathrm{e})$.

All saddle-node bifurcations (under mild technical assumptions) are locally topologically conjugate to the normal form mapping [206]

$$y_{n+1} = \nu + y_n + y_n^2 =: g_{\mathrm{sn}}(y, \nu). \tag{6.33}$$

One can easily compute that a saddle-node bifurcation takes place in g_{sn} at $(y, \nu) = (0, 0)$ with branches of equilibria given by $y^*_\pm(\nu) = \sqrt{-\nu}$ existing for all $\nu \leq 0$. A conjugacy between the saddle-node bifurcation in (6.29) and the normal form g_{sn} comes in the form of a parameter-dependent transformation $(y, \nu) = h(x, \mu)$ that is invertible on a neighborhood of the saddle-node bifurcation point. The precise structure of these transformations is system dependent and typically obtained through nonconstructive means, such as the implicit function theorem, meaning the precise link between (6.29) and the normal form g_{sn} is unknown. Thus, we employ an autoencoder to approximate this transformation, specifically with the encoder η approximating h and the decoder ρ approximating its inverse.[6] We shift the variables so that the bifurcation in (6.29) at $(x, \mu) = (1, \mathrm{e})$ takes place at the origin. The loss function for training the network uses the standard autoencoder loss (6.23), but we further include an additional term,

$$\mathcal{L}_4 = \|\eta(0, 0)\|_2^2, \tag{6.34}$$

which helps to enforce that the bifurcation point in the training data $(0, 0)$ is mapped to the bifurcation point in the normal form $\mathrm{g_{sn}}$. Training data is obtained by sampling 5000 points from a square in (x, μ)-space centered at the bifurcation point with side length 1 in x and 0.4 in μ. We find that a step of $S = 1$ in the loss function is sufficient to obtain promising results.

[6] In [137] the authors seek two separate autoencoders to move from (x, μ) to the normal form coordinates (y, ν). They use one to transform x to y and another to transform μ to ν. Although the reported results appear positive, such a disentangling of the state variable from the parameter does not seem to have a basis in theory. In fact, proofs of topological conjugacy with normal forms requires the conjugacy to be a function of both (x, μ) together. Since this disentanglement is not justified rigorously, we use only one encoder and decoder which depend on both the state and the parameter in order to provide an analytical justification for our computational implementation.

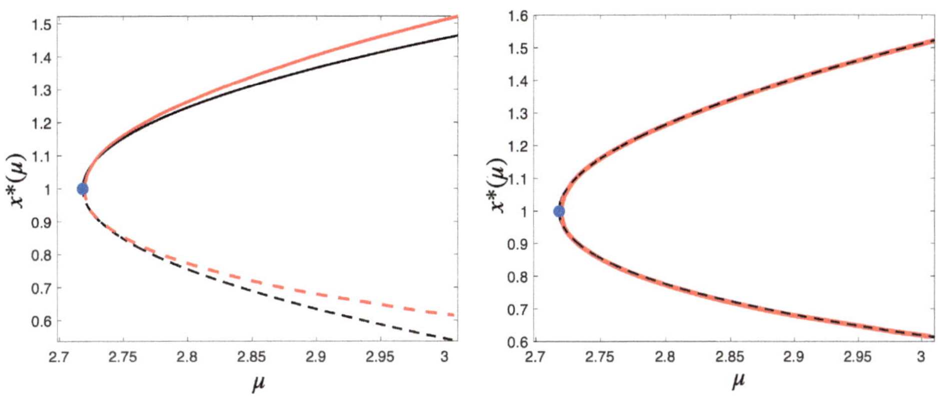

Figure 6.4. *Left: The bifurcation diagram of* (6.29) *obtained through the autoencoder (red) and the Taylor expansion approximation (black) given in* (6.32). *The autoencoder data are obtained by applying the decoder* ρ *to the branches of equilibria* $\pm\sqrt{-\nu}$ *of the normal form* (6.33). *Stable branches are solid, while unstable branches are dashed. Right: The bifurcation branches from the autoencoder (red) against the numerically obtained fixed points of* (6.29) *(black). In both panels the bifurcation point* $(x,\mu) = (1, \mathrm{e})$ *is given as a blue dot.*

To demonstrate the validity of a trained network, we use the normal form to produce the bifurcation diagram of (6.29) near the saddle-node bifurcation at $(x,\mu) = (1, \mathrm{e})$. Precisely, since fixed points are mapped into each other by conjugacies, we compute values along the curves of fixed points $(\pm\sqrt{-\nu}, \nu)$ to the normal form (6.33) and then apply the trained decoder ρ to map them to the approximate location of the curves of fixed points in (x,μ)-space. In Figure 6.4 we present the results of this process. In the left panel we compare the locations of these branches as x against μ with the Taylor expansions (6.32), where we see the tangency of the Taylor approximation as one approaches the bifurcation point. We further include the stability of the branches since $-\sqrt{-\nu}$ is the stable branch in g_{sn}, and its image under the decoder $\rho(-\sqrt{-\nu}, \nu)$ is the stable branch $x_+^*(\mu)$ in (6.29). The same is true for the unstable branch. As a final check, we use a root-finding algorithm to numerically obtain the fixed points of (6.29) and compare them to the branches obtained by the neural network. The results are plotted in the right panel of Figure 6.4, with the curves being nearly indistinguishable to plotting accuracy.

Beyond saddle-node bifurcations, there are three other generic bifurcations that can take place in one-dimensional and one-parameter maps, i.e., $(d,p) = (1,1)$. All of these bifurcations are characterized by $|f_x(x^*, \mu^*)| = 1$ and caused by varying μ in a neighborhood of μ^*. Beyond the saddle-node bifurcation of Example 6.5, we have the following (under fairly general conditions):

- A **transcritical bifurcation** takes place when f has two fixed points, one stable and one unstable, for μ to the left and right of μ^*, but these fixed points coincide at the bifurcation point (x^*, μ^*). As μ crosses μ^*, the stability of each branch of fixed points flips. Such a bifurcation is locally topologically equivalent to the normal form

$$y_{n+1} = y_n + y_n(\nu - y_n) =: g_{\mathrm{tr}}(y_n, \nu). \tag{6.35}$$

- A **pitchfork bifurcation** takes place when the number of fixed points of f changes (locally) from one to three as one crosses μ^*. The locally equivalent normal form is

$$y_{n+1} = y_n + y_n(\nu + r y_n^2) =: g_{\mathrm{pf}}(y_n, \nu), \tag{6.36}$$

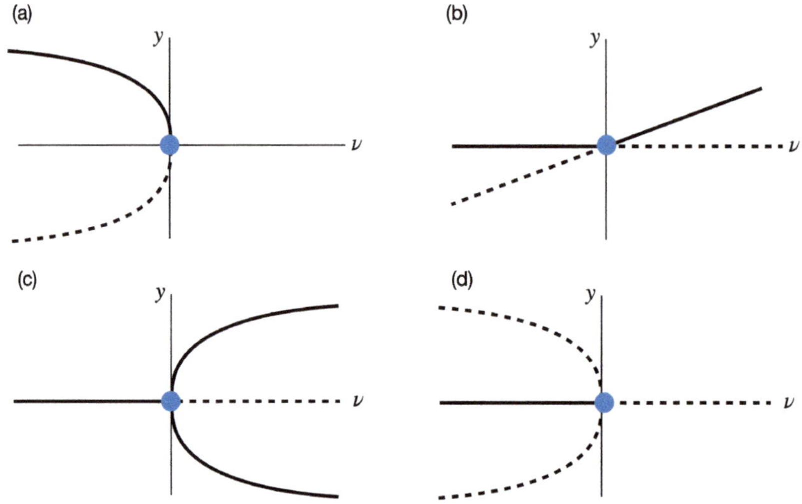

Figure 6.5. *Bifurcation diagrams for the normal forms discussed in this section. Thick lines are solution branches of fixed points, and the bifurcation point* $(\nu, y) = (0, 0)$ *is represented by a blue dot. Solid branches are stable, and dashed branches are unstable.* (a) *Saddle-node bifurcation, as governed by the normal form* (6.33). (b) *Transcritical bifurcation, as governed by* (6.35). (c) *Supercritical pitchfork bifurcation, as governed by* (6.36) *with* $r = -1$. (d) *Subcritical pitchfork bifurcation, as governed by* (6.36) *with* $r = 1$. *Super- and subcritical period-doubling bifurcation diagrams, as governed by* (6.37), *are the same as* (c)–(d) *except that the parabolic bifurcation curves are elements of a* 2*-cycle instead of fixed points.*

where $r = \pm 1$ is determined by the stability of the emerging branches. In both cases the bifurcation is brought on from the trivial branch $y = 0$ moving from stable to unstable as ν crosses 0 from left to right. If $r = -1$, the bifurcation is **supercritical**, with fixed points $y^*_\pm(\nu) = \pm\sqrt{\nu}$ emerging as stable branches of fixed points from the bifurcation point $(y^*, \nu^*) = (0, 0)$. If $r = 1$, the bifurcation is **subcritical**, and the branches of fixed points $y^*_\pm(\nu) = \pm\sqrt{-\nu}$ are unstable and collapse onto $y = 0$ at the bifurcation point.

- A **period-doubling bifurcation** takes place when a fixed point changes stability and a 2-cycle emerges. The locally topologically equivalent normal form is given by

$$y_{n+1} = -y_n - y_n(\nu + r y_n^2) =: g_{\text{pd}}(y_n, \nu), \tag{6.37}$$

where again $r = \pm 1$. For both values of r the trivial branch $y = 0$ moves from stable to unstable as ν crosses 0 from left to right. At the bifurcation point $(y^*, \nu^*) = (0, 0)$ a 2-cycle given by $(y_1^*(\nu), y_2^*(\nu)) = (\sqrt{-r\nu}, -\sqrt{-r\nu})$ emerges and exists for $r\nu \leq 0$. The 2-cycle is stable when $r = -1$ (**supercritical**) and unstable when $r = 1$ (**subcritical**).

Bifurcation diagrams for each of the normal forms above are sketched in Figure 6.5. The reader is referred to [206, 109, 162, 97] for more in-depth discussions of these bifurcations and in particular the results related to topological conjugacies.

These bifurcations are often found in higher-dimensional mappings as well. In order to transform bifurcations in maps in dimensions $d \geq 2$ to an appropriate normal form we look to the center manifold theorem [53]. Going back to our assumptions that the linearization $f_x(x^*, \mu^*)$ has exactly one eigenvalue at ± 1, we further assume that there are exactly d_s eigenvalues inside the unit circle and d_u eigenvalues outside, so $d = 1 + d_s + d_u$. Then the dynamics of

$x_{n+1} = f(x_n, \mu)$ in a neighborhood of (x^*, μ^*) are topologically conjugate to the map [162]

$$\begin{aligned} [y_c]_{n+1} &= g([y_c]_n, \mu), \\ [y_s]_{n+1} &= A_s[y_s]_n, \\ [y_u]_{n+1} &= A_u[y_u]_n, \end{aligned} \tag{6.38}$$

where $(y_c, y_s, y_u) \in \mathbb{R}^1 \times \mathbb{R}^{d_s} \times \mathbb{R}^{d_u}$, A_s is a real-valued matrix with all eigenvalues inside the unit circle, and A_u is a matrix with all eigenvalues outside the unit circle. The component y_c constitutes the one-dimensional component on the center manifold, which we see has now been decoupled from the stable y_s and unstable y_u components. Thus, we may apply a normal form transformation to further conjugate $g(y_c, \mu)$ into one of the normal forms presented above.

The matrices A_s and A_u in (6.38) can be chosen as the Jordan normal form of a matrix with eigenvalues given by those of $f_x(x^*, \mu^*)$ inside and outside the unit circle, respectively. Further, a more universal form is presented in [162, Chapter 5] which goes back to results from Shoshitaishvili [255] and is as follows. When $\det(A_s) > 0$, the dynamics of the stable component $[y_s]_{n+1} = A_s[y_s]_n$ are conjugate to the *standard orientation-preserving stable node*

$$z_{n+1} = \frac{1}{2} z_n, \tag{6.39}$$

where $z \in \mathbb{R}^{d_s}$. Alternatively, when $\det(A_s) < 0$, the dynamics of the stable component are conjugate to the *standard orientation-reversing stable node*

$$z_{n+1} = \begin{bmatrix} \frac{1}{2} & 0 & \cdots & 0 & 0 \\ 0 & \frac{1}{2} & \cdots & 0 & 0 \\ \vdots & \vdots & \ddots & \vdots & \vdots \\ 0 & 0 & \cdots & \frac{1}{2} & 0 \\ 0 & 0 & \cdots & 0 & -\frac{1}{2} \end{bmatrix} z_n. \tag{6.40}$$

The unstable components are similarly topologically conjugate to the *standard unstable node*, which is defined similarly to the standard stable node but with 2 taking the place of the $\frac{1}{2}$. Together the standard stable and unstable nodes define the *standard saddle map* on $\mathbb{R}^{d_s+d_u}$.

Example 6.6. In this example we will implement a change of coordinates near a period-doubling bifurcation in a three-dimensional map using the center manifold theorem. To this end, we investigate a simplified version of the adaptive control system from [99], given by

$$\begin{bmatrix} x_1 \\ x_2 \\ x_3 \end{bmatrix}_{n+1} = \begin{bmatrix} x_2 \\ 1 + \mu x_1 + x_2 x_3 \\ x_3 - \mu x_1 x_2 - x_2^2 x_3 \end{bmatrix}_n, \tag{6.41}$$

with $\mu \in \mathbb{R}$ being the bifurcation parameter.

First, for each μ there is exactly one fixed point, given by $(x_1, x_2, x_3) = (1, 1, -\mu)$. Linearizing (6.41) about this fixed point results in the matrix

$$f_x((1, 1, -\mu), \mu) = \begin{bmatrix} 0 & 1 & 0 \\ \mu & -\mu & 1 \\ -\mu & \mu & 0 \end{bmatrix}, \tag{6.42}$$

which has characteristic polynomial $\chi(\zeta) = \zeta^3 + \mu\zeta^2 - 2\mu\zeta + \mu$. Since we are interested in period-doubling bifurcations, we seek parameter values for which $\zeta = -1$ is an eigenvalue of the linearization $f_x((1, 1, -\mu), \mu)$. Thus,

$$\chi(-1) = 0 \iff -1 + \mu + 2\mu + \mu = 0 \iff \mu = \frac{1}{4}. \tag{6.43}$$

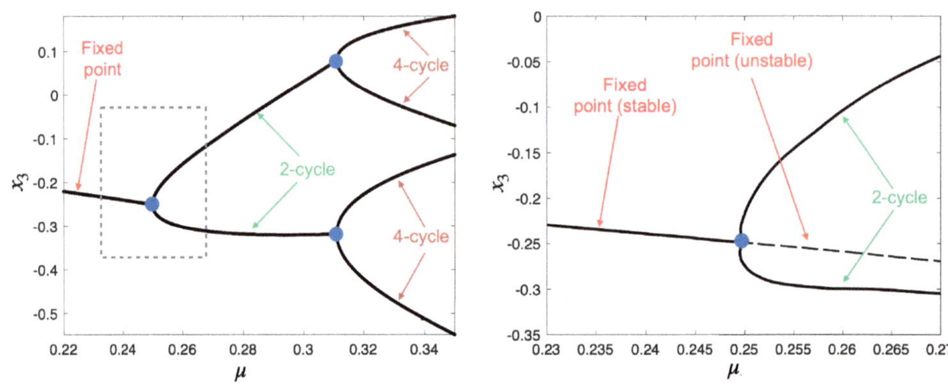

Figure 6.6. *Left: The bifurcation diagram of* (6.41) *in the* (μ, x_3)*-plane obtained by iterating random initial conditions and plotting the limiting attractor over a range of* μ *values. In the presented parameter range there are two period-doubling bifurcations (blue dots), the first of which takes place at* $\mu = 1/4$. *Right: The bifurcation diagram of* (6.41) *produced using the normal form* (6.46) *relating to the boxed portion of the diagram in the left panel.*

At $\mu = \mu^* := 1/4$ the roots of the characteristic polynomial are given by $\zeta = -1, (3 \pm \sqrt{7}\mathrm{i})/8$, giving $d_s = 2$ and $d_u = 0$ for the application of the center manifold theorem. Since $\zeta = -1$ is an isolated root of the characteristic equation at $\mu = 1/4$, there exists a smooth, real-valued curve $\zeta(\mu)$ that exists for μ near $1/4$, satisfies $\zeta(1/4) = -1$, and solves $\chi(\zeta(\mu)) = 0$ for all μ in its domain. Differentiating the expression $\chi(\zeta(\mu)) = 0$ with respect to μ gives

$$3\zeta(\mu)^2\zeta'(\mu) + \zeta(\mu)^2 + 2\mu\zeta(\mu)\zeta'(\mu) - 2\zeta(\mu) - 2\mu\zeta'(\mu) + 1 = 0, \tag{6.44}$$

and evaluating at $\mu = 1/4$ gives

$$\zeta'(1/4) = -2. \tag{6.45}$$

Hence, $\zeta(\mu)$ is decreasing at $\mu = 1/4$, meaning that the fixed point $(x_1, x_2, x_3) = (1, 1, -\mu)$ is stable in an interval to the left of $\mu = 1/4$ and unstable in an interval to the right.

The loss of stability of the fixed point as μ crosses $1/4$ from left to right can be seen in Figure 6.6, where we provide the location of the attractor in the (μ, x_3)-plane over a range of μ values. Importantly, this bifurcation diagram shows that at $\mu = 1/4$ we indeed have a period-doubling bifurcation that births a stable 2-cycle, thus giving that the bifurcation is supercritical. Furthermore, this 2-cycle undergoes another period-doubling bifurcation, resulting in a stable 4-cycle. Of course this second period-doubling bifurcation is not captured by the normal form (6.37), and so any normal form transformation must be sufficiently localized about the period-doubling bifurcation at $\mu = 1/4$ to exclude this secondary doubling.

As in the previous example, we will shift the bifurcation point $(x^*, \mu^*) = (1, 1, -1/4, 1/4)$ to the origin to produce the network training data. Data is gathered by randomizing 10^4 points in a rectangle with side lengths 0.4 in the x_1, x_2, x_3 coordinates and side length 0.04 in the parameter μ. Since $d_s = 2$ and $d_u = 0$, we seek a local conjugacy with the map

$$\begin{bmatrix} y_1 \\ y_2 \\ y_3 \end{bmatrix}_{n+1} = \begin{bmatrix} -y_1 - y_1(\nu - y_1^2) \\ \frac{3}{8}y_2 + \frac{\sqrt{7}}{8}y_3 \\ -\frac{\sqrt{7}}{8}y_2 + \frac{3}{8}y_3 \end{bmatrix}_n . \tag{6.46}$$

In this case we see that the center component, y_1, takes the form of a supercritical period-doubling bifurcation, while the stable components, y_2 and y_3, are linear and evolve according to matrix multiplication via the Jordan normal form associated to eigenvalues $(3 \pm \sqrt{7}\mathrm{i})/8$.

The right panel of Figure 6.6 provides the bifurcation diagram in the (μ, x_3)-plane produced by the center manifold equation (6.46) and the learned autoencoder coordinate transformation. Notice that through the normal form we can track the steady state $(1,1,-\mu)$ beyond the bifurcation point where it becomes unstable. This is because the destabilized fixed point is given by $(y_1, y_2, y_3) = (0,0,0)$ for all ν in (6.46), and so the learned decoder ρ approximately maps the origin in the y variables over a range of ν back to the fixed point for the corresponding values of μ. Of course the location of the unstable fixed point is known for all values of μ simply by analyzing the map (6.41); however, in principle such normal form transformations allow one to track destabilized states even when their location in (x, μ) space is unknown.

One may similarly learn a transformation of the stable components y_2 and y_3 into the form (6.39) instead of the Jordan normal form used in (6.46). Results are similar to those reported above but are not presented to avoid repetition. The reader is encouraged to perform their own experiments that compare and contrast these two targeted forms for the dynamics.

Throughout this section we have dealt with the simple bifurcation scenario that only one eigenvalue of $f_x(x^*, \mu^*)$ lies on the unit circle. Naturally, all of the work above extends to the case when there are $d_c \geq 1$ eigenvalues on the unit circle, with the conjugate center manifold equation $[y_c]_{n+1} = g([y_c]_n, \mu)$ now a mapping in d_c variables. The resulting bifurcations due to multiple eigenvalues crossing the unit circle at the same time lead to significantly more complex normal forms that are difficult to analyze. Moreover, many of these normal forms have not been proven to be topologically conjugate to the original center manifold dynamics. The reason for this is that the truncated normal forms produce dynamics that are not structurally stable, such as homo- and heteroclinic orbits, meaning that the inclusion of more terms in a Taylor series approximation could destroy these structures and prevent one from obtaining topologically conjugacies with simple, low-degree mappings akin to those discussed in this section.

Finally, the inclusion of higher-order terms in a normal form is not unique to having more than one eigenvalue on the unit circle. For example, [97] demonstrates that if a smooth conjugacy between a one-dimensional center manifold mapping and a normal form is desired, then certain higher-order terms in the normal form must be included. Such a result likely has important implications for future results that demonstrate the ability of neural networks to sufficiently approximate the conjugacy between local bifurcations and their normal forms since activation functions are typically smooth (almost everywhere), resulting in smooth encoders and decoders.

6.4 ▪ Global Linearization

In Example 6.2 we saw that conjugacies can be used to identify local coordinate transformations from a nonlinear system to a linear one. Practically, identifying such a conjugacy does very little for understanding the dynamics of a nonlinear system since one is restricted to a small neighborhood of a hyperbolic fixed point. However, the Koopman operator provides an alternative formulation of the system that offers a *global* linearization of the dynamics. The linearity of the Koopman operator means that its dynamics can be decomposed into linear combinations of its eigenfunctions. Recall that in Section 2.4 we proposed the method of EDMD to approximate the Koopman eigenfunctions as linear combinations of a given dictionary of functions. In this section we will provide an alternative, and potentially more robust, method of learning the Koopman eigenfunctions using neural networks, as proposed in the works [180, 95].

We begin with the usual setting of a discrete dynamical system $x_{n+1} = f(x_n)$, with $f : \mathbb{R}^d \to \mathbb{R}^d$. Recall that a Koopman eigenfunction, $\psi : \mathbb{R}^d \to \mathbb{C}$, with eigenvalue $\mu \in \mathbb{C}$, satisfies

$$\psi \circ f = \mu\psi. \tag{6.47}$$

Dynamically this means that $\psi(x_{n+1}) = \mu\psi(x_n)$ for any sequence of iterates $\{x_0, x_1, x_2, \dots,\}$

of the map f, and so $\psi(x_n) = \mu^n\psi(x_0)$. Notice that if we define $z_n = \psi(x_n) \in \mathbb{C}$, we arrive at the linear discrete dynamical system $z_{n+1} = \mu z_n$, with initial condition $z_0 = \phi(x_0)$. Thus, ψ represents a global linearizing change of variable for the nonlinear system $x_{n+1} = f(x_n)$.

We can take this further, considering a set of $r \geq 1$ Koopman eigenfunctions $\{\psi_1, \psi_2, \dots, \psi_r\}$ with corresponding eigenvalues $(\mu_1, \mu_2, \dots, \mu_r)$. This leads to the linear dynamical system

$$\begin{bmatrix} z_1 \\ z_2 \\ \vdots \\ z_r \end{bmatrix}_{n+1} = \begin{bmatrix} \mu_1 & 0 & \cdots & 0 \\ 0 & \mu_2 & \cdots & 0 \\ \vdots & \vdots & \ddots & \vdots \\ 0 & 0 & \cdots & \mu_r \end{bmatrix} \begin{bmatrix} z_1 \\ z_2 \\ \vdots \\ z_r \end{bmatrix}_n , \tag{6.48}$$

which is diagonal in the complex-valued coordinates $z_1 = \phi_1(x), \dots, z_r = \phi_r(x)$. Since neural networks produce real-valued outputs, we separate out real and imaginary parts as $z_j = y_{2j-1} + \mathrm{i}y_{2j}$ and $\mu_j = \alpha_j + \mathrm{i}\beta_j$ to write (6.48) as the $2r$-dimensional block-diagonal system

$$\begin{bmatrix} y_1 \\ y_2 \\ y_3 \\ y_4 \\ \vdots \\ y_{2r-1} \\ y_{2r} \end{bmatrix}_{n+1} = \begin{bmatrix} \alpha_1 & \beta_1 & 0 & 0 & \cdots & 0 & 0 \\ -\beta_1 & \alpha_1 & 0 & 0 & \cdots & 0 & 0 \\ 0 & 0 & \alpha_2 & \beta_2 & \cdots & 0 & 0 \\ 0 & 0 & -\beta_2 & \alpha_2 & \cdots & 0 & 0 \\ \vdots & \vdots & \vdots & \vdots & \ddots & \vdots & \vdots \\ 0 & 0 & 0 & 0 & \cdots & \alpha_r & \beta_r \\ 0 & 0 & 0 & 0 & \cdots & -\beta_r & \alpha_r \end{bmatrix} \begin{bmatrix} y_1 \\ y_2 \\ y_3 \\ y_4 \\ \vdots \\ y_{2r-1} \\ y_{2r} \end{bmatrix}_n , \tag{6.49}$$

with real variables $y_{2j-1} = \mathrm{Re}(\psi_j(x))$ and $y_{2j} = \mathrm{Im}(\psi_j(x))$ coming from the Koopman eigenfunctions. Thus, considering a fixed set of Koopman eigenvalue-eigenfunction pairs $\{(\mu_j, \psi_j)\}_{j=1}^r$ we can take $g : \mathbb{R}^r \to \mathbb{R}^r$ to represent the linear mapping (6.49). The change of variable

$$\Psi(x) := (\mathrm{Re}(\psi_1(x)), \mathrm{Im}(\psi_1(x)), \dots, \mathrm{Re}(\psi_r(x)), \mathrm{Im}(\psi_r(x)))^\intercal \tag{6.50}$$

satisfies the conjugacy relationship $\Psi \circ f = g \circ \Psi$. However, Ψ is not necessarily invertible, and so the relationship between f and g is not a true conjugacy.

The question now becomes, which Koopman eigenfunctions do we use to build the linear map g? First note that the change of variable Ψ in (6.50) does indeed provide a new coordinate system in which the dynamics are linear; however, it does not necessarily provide a method to return to the original domain of f. To overcome this we strive to use Koopman eigenfunctions so that the state-space observables belong to their span, meaning we can write

$$x = \sum_{j=1}^{r} a_j \mathrm{Re}(\psi_j(x)) + b_j \mathrm{Im}(\psi_j(x)) \tag{6.51}$$

for some vectors $a_j, b_j \in \mathbb{R}^d$. Then we have a way of moving back from the linear coordinate system since this relationship guarantees that the function $\Phi : \mathbb{R}^{2r} \to \mathbb{R}^d$ defined by

$$\Phi(y) = \sum_{j=1}^{r} a_j y_{2j-1} + b_j y_{2j} \tag{6.52}$$

is such that $\Phi(\Psi(x)) = x$ for every $x \in \mathbb{R}^d$. However, we should expect that $r > d$, and so Φ is only a left inverse of Ψ, as the spaces $\mathbb{R}^d$ and $\mathbb{R}^r$ are not homeomorphic when $d \neq r$. Nonetheless, it does bring us back from the linear coordinate system to the original state-space, which is all that we desire to identify Koopman eigenfunctions with an autoencoder.

To employ an autoencoder to discover a global linearization of the system $x_{n+1} = f(x_n)$, we will treat $r \geq 1$ as a hyperparameter to be fixed before the training process, much like the widths and number of layers in the encoder and decoder neural networks. Then the encoder $\eta : \mathbb{R}^d \to \mathbb{R}^r$ is trained to replicate Ψ in (6.50), while the decoder $\rho : \mathbb{R}^r \to \mathbb{R}^d$ is completely linear and trained to replicate its inverse Φ. The real and imaginary parts of the eigenvalues (α_j, β_j) are left as trainable parameters for the neural network. The idea is that each component of the output of the encoder η is trained to approximate an eigenfunction of the Koopman operator so that the state-space observables lie in their linear span to produce the decoder ρ. The (α_j, β_j) are then identified along with this optimal basis of eigenfunctions.

Example 6.7. We now examine the performance of using autoencoders to discover global linearizing transformations using Koopman eigenfunctions. We return to the simple two-dimensional nonlinear system from Example 2.7. Precisely, we consider

$$\begin{bmatrix} x_1 \\ x_2 \end{bmatrix}_{n+1} = \begin{bmatrix} \lambda_1 x_1 \\ \lambda_2(x_2 - x_1^2) \end{bmatrix}_n, \qquad \lambda_1, \lambda_2 \in \mathbb{R}. \tag{6.53}$$

The function $\Psi : \mathbb{R}^2 \to \mathbb{R}^3$ is given by the previously identified Koopman eigenfunctions

$$\Psi(x_1, x_2) = \left(x_1, x_2 - \frac{\lambda_2}{\lambda_2 - \lambda_1^2} x_1^2, x_1^2 \right), \tag{6.54}$$

which transforms (6.53) to the diagonal linear system

$$\begin{bmatrix} y_1 \\ y_2 \\ y_3 \end{bmatrix}_{n+1} = \begin{bmatrix} \lambda_1 & 0 & 0 \\ 0 & \lambda_2 & 0 \\ 0 & 0 & \lambda_1^2 \end{bmatrix} \cdot \begin{bmatrix} y_1 \\ y_2 \\ y_3 \end{bmatrix}_n \tag{6.55}$$

on setting $(y_1, y_2, y_3,) = \Psi(x_1, x_2)$. Furthermore, we can recover the state-space observables x_1 and x_2 through the linear transformation $\Phi : \mathbb{R}^3 \to \mathbb{R}^2$ by

$$\Phi(y_1, y_2, y_3) = \begin{bmatrix} y_1 \\ y_2 + \frac{\lambda_2}{\lambda_2 - \lambda_1^2} y_3 \end{bmatrix} \tag{6.56}$$

since $\Phi(\Psi(x_1, x_2)) = (x_1, x_2)$.

We now seek to identify the system (6.55) using an autoencoder. Throughout this example we fix $\lambda_1 = 1.5$ and $\lambda_2 = 0.5$; however, different values lead to nearly identical results. First we will gather $K = 5000$ data points drawn uniformly from the square $[-2, 2]^2$ to train the network on. We fix $r = 3$ since we know that we only require three Koopman eigenfunctions to linearize the system; however, for other examples r must be treated as a hyperparameter and typically has to be large. Initially we fix the conjugate linear system to be

$$g(y_1, y_2, y_3) = (1.5y_1, 0.5y_2, 2.25y_3)^\intercal, \tag{6.57}$$

thus fixing the Koopman eigenvalues so they are not learned during the training process. This allows us to first examine the discovery of Koopman eigenfunctions.

The resulting Koopman eigenfunctions from the trained network with fixed eigenvalues are plotted in Figure 6.7(a)-(c). Precisely, the eigenfunction corresponding to the eigenvalue $\lambda_1 = 1.5$ is given in panel (a), where it appears to display only linear variation in the x_1 direction. However, since yellow in the contour plots represents larger values, it appears that the discovered eigenfunction is $-Cx_1$ for some $C > 0$. This is consistent with the fact that eigenfunctions are

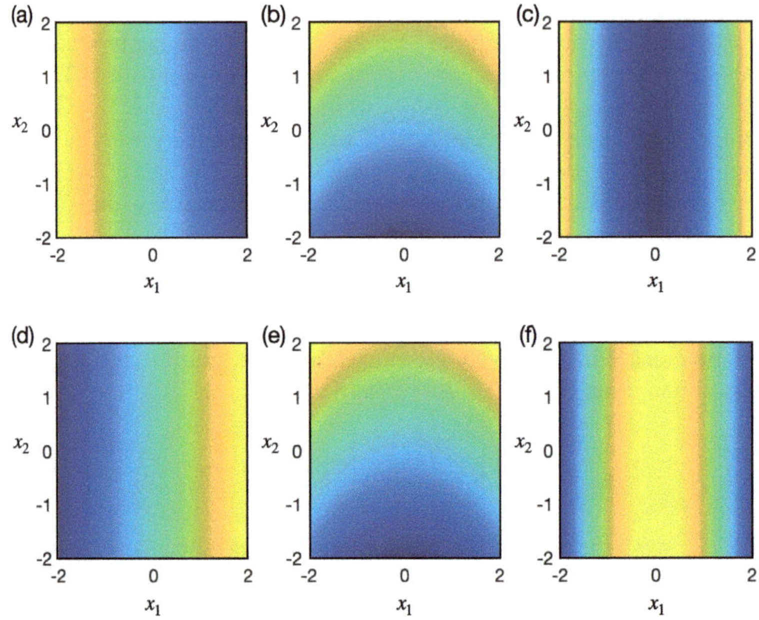

Figure 6.7. *Contour plots of Koopman eigenfunctions to linearize the map* (6.53) *with* $\lambda_1 = 1.5$ *and* $\lambda_2 = 0.5$. *Yellow denotes larger values, while blue denotes those that are smaller.* (a)–(c) *are the discovered eigenfunctions with the eigenvalues* (a) λ_1, (b) λ_2, *and* (c) λ_1^2 *fixed during the training process, while* (d)–(f) *are the discovered eigenfunctions with the corresponding eigenvalues discovered during the neural network training process.*

unique up to scalar multiplication. Similarly, the contour plot in panel (b) corresponds approximately to a positive scalar multiple of the Koopman eigenfunction $\psi_2(x_1, x_2) = x_2 - \lambda_2/(\lambda_2 - \lambda_1^2)x_1^2$, and that in panel (c) approximately corresponds to a multiple of $\psi_3(x_1, x_2) = x_1^2$.

We can further check these results by examining the trained decoder $\rho : \mathbb{R}^3 \to \mathbb{R}^2$. In this case it approximates $\Phi(y_1, y_2, y_3)$ by

$$\rho(y_1, y_2, y_3) = \begin{bmatrix} \texttt{-1.55041}y_1 + \texttt{0.00373}y_2 + \texttt{0.03603}y_3 \\ \texttt{0.00248}y_1 + \texttt{1.18096}y_2 - \texttt{0.66994}y_3 \end{bmatrix}. \tag{6.58}$$

From these results we get an approximation of the true inversion operator Φ with the same approximate sparsity pattern but different constants. This is because the encoder $\eta : \mathbb{R}^2 \to \mathbb{R}^3$ is discovering different multiples of the eigenfunctions, and the decoder compensates for it.

Now let us turn to the case when the Koopman eigenvalues are trained by the network as well. Thus, the conjugate linear map will be given by

$$g(y_1, y_2, y_3) = (\mu_1 y_1, \mu_2 y_2, \mu_3 y_3)^\intercal, \tag{6.59}$$

where $(\mu_1, \mu_2, \mu_3) \in \mathbb{R}^3$ are set as variables in the network to be identified during training. The trained autoencoder network returns

$$(\mu_1, \mu_2, \mu_3) = (\texttt{2.24662}, \texttt{0.49965}, \texttt{1.499636}), \tag{6.60}$$

all of which display a relative error of less than 1% of the true Koopman eigenvalues. However, we note that this highlights a further degeneracy in the discovery of global linearizing transforms since the discovered eigenvalues can come as any permutation of the true eigenvalues λ_1, λ_2, and

λ_1^2. Figure 6.7(d)–(f) again provides contour plots of the discovered eigenfunctions, ordered in terms of the eigenvalues λ_1, λ_2, and λ_1^2, respectively, for comparison with the previous trial.

The previous experiments were undertaken with the assumption that the eigenvalues are real. In practice, this is not something that can be assumed, and so we should work with the augmented system (6.49), which accounts for real and imaginary parts of both the eigenvalues and eigenfunctions. Thus, we repeat the experiment with trainable (α_j, β_j), $j = 1, 2, 3$, to construct linear $g : \mathbb{R}^6 \to \mathbb{R}^6$ of the form (6.49). The trained autoencoder returns Koopman eigenvalues

$$\begin{aligned} \mu_1 &= \alpha_1 + \mathrm{i}\beta_1 = \mathtt{1.49908} - \mathtt{0.00576}\mathrm{i}, \\ \mu_2 &= \alpha_2 + \mathrm{i}\beta_2 = \mathtt{0.49936} - \mathtt{0.00123}\mathrm{i}, \\ \mu_3 &= \alpha_3 + \mathrm{i}\beta_3 = \mathtt{2.26499} - \mathtt{0.01593}\mathrm{i}, \end{aligned} \tag{6.61}$$

again agreeing with a relative error of less than 0.5% of the true Koopman eigenvalues. However, a curious degeneracy arises in this situation. Since the imaginary parts are nearly zero, the system completely diagonalizes into observables $y_1, \dots, y_6$ that only have to sum up to recover the state-space observables via the decoder $\rho : \mathbb{R}^6 \to \mathbb{R}^2$. Thus, it is not necessary that y_1, y_3, y_5 be the real parts of the Koopman eigenfunctions with corresponding imaginary parts y_2, y_4, y_6, respectively. For example, we have that the approximate eigenfunctions y_1 and y_2 are both learned to approximate the true eigenfunction $\psi_1(x_1, x_2) = x_1$. Further evidence for this comes from examining the first component of the learned decoder function,

$$[\rho(y_1, y_2, y_3, y_4, y_5, y_6)]_1 = \mathtt{0.12328}y_1 - \mathtt{0.17708}y_2 + \dots, \tag{6.62}$$

where we can see a nontrivial contribution to reconstructing x_1 from both the "real part" y_1 and the "imaginary part" y_2. The same thing is happening with the other variables, all stemming from the degeneracy of doubling the system size but learning that all imaginary contributions are zero.

It is important to note that in the previous example we knew the value of r to produce the linearized Koopman coordinate system. In all but the simplest examples, we do not have access to the Koopman eigenfunctions, and therefore we cannot determine which of them can be used to reconstruct the state-space observables to build the inverse $\Psi : \mathbb{R}^{2r} \to \mathbb{R}^d$. Thus, r must be treated as a hyperparameter of the neural network, with the maxim that the larger the value of r, the better the ability of the eigenfunctions to build Φ. For example, in [95] a value of $r = 21$ is used to produce linearizing observables for Burgers equation (2.68), while values up to $r = 128$ are used for the Kuramoto–Sivashinsky equation.

For the remainder of this section we focus on a special case where the value of r is known explicitly. We consider a smooth **Hamiltonian** function $\mathcal{H} : \mathbb{R}^{2d} \to \mathbb{R}$. In phase space coordinates $(p, q) \in \mathbb{R}^d \times \mathbb{R}^d$, the Hamiltonian function leads to Hamilton's equations:

$$\begin{aligned} \dot{q} &= \frac{\partial \mathcal{H}}{\partial p}(p, q), \\ \dot{p} &= -\frac{\partial \mathcal{H}}{\partial q}(p, q). \end{aligned} \tag{6.63}$$

Here p is referred to as the generalized momentum and q the generalized position. One can check that the Hamiltonian $\mathcal{H}$ is a conserved quantity for (6.63), as introduced in Section 3.5, giving that solutions of (6.63) are restricted to the generically $(2d - 1)$-dimensional level sets of $\mathcal{H}$.

As we saw in Example 3.12, it is possible for (6.63) to have more conserved quantities beyond the Hamiltonian, which in turn would restrict the flow of the system to lower-dimensional manifolds embedded in the $2d$-dimensional phase space. The system (6.63) is said to be **integrable** if there exist d independent functions $f_1,\dots,f_d:\mathbb{R}^{2d}\to\mathbb{R}$ such that $\{f_j,f_k\}=0$ for all $j,k=1,\dots,d$, where $\{\cdot,\cdot\}$ is the **Poisson bracket**, given by

$$\{f,g\}=\sum_{j=1}^{d}\frac{\partial f}{\partial q_j}\frac{\partial g}{\partial p_j}-\frac{\partial f}{\partial p_j}\frac{\partial g}{\partial q_j}. \tag{6.64}$$

From the above we know that one such conserved quantity is the Hamiltonian function. Furthermore, from the fact that $\{f_j,f_k\}=0$ for all $j,k=1,\dots,d$ we can see that each f_j is constant along trajectories of (6.63) as well. Indeed, given $(p(t),q(t))$ a solution of (6.63), we find that

$$\begin{aligned}\frac{\mathrm{d}}{\mathrm{d}t}f_j(p(t),q(t))&=\frac{\partial f_j}{\partial q}\cdot\dot q+\frac{\partial f_j}{\partial p}\cdot\dot p\\&=\frac{\partial f_j}{\partial q}\cdot\frac{\partial\mathcal{H}}{\partial p}-\frac{\partial f_j}{\partial p}\cdot\frac{\partial\mathcal{H}}{\partial q}\\&=\{f_j,\mathcal{H}\}\\&=0\end{aligned} \tag{6.65}$$

for all $j=1,\dots,d$, showing that the f_j are conserved quantities of (6.63). Thus, trajectories of integrable systems are confined to the intersection of level sets of the functions f_j.

The Liouville–Arnold theorem [18] states that with $c_j\in\mathbb{R}$ fixed, if

$$\mathcal{M}_f:=\{(p,q)\in\mathbb{R}^{2d}:\ f_j(p,q)=c_j,\ j=1,\dots,d\} \tag{6.66}$$

is compact and connected, then $\mathcal{M}_f$ defines an invariant torus for the system (6.63). Moreover, one can introduce the **action-angle** coordinates $I_1,\dots,I_d,\theta_1,\dots,\theta_d$ with $\theta_j(p,q)$ defining the angular coordinates on $\mathcal{M}_f$, and the **actions** $I_j=I_j(p,q)$ are constants of the motion of the system. Importantly, the transformation from (p,q) in (6.63) to (I,θ) is invertible and satisfies

$$\{I_j,I_k\}=0,\quad \{\theta_j,\theta_k\}=0,\quad \{\theta_j,I_k\}=\begin{cases}1, & j=k,\\ 0, & j\neq k.\end{cases} \tag{6.67}$$

In these action-angle coordinates the equations of motion (6.63) become

$$\begin{aligned}\dot I_k&=0,\\ \dot\theta_k&=\omega_k(I_1,\dots,I_d)\end{aligned} \tag{6.68}$$

for each $k=1,\dots,d$. System (6.68) is linear and exactly solvable, yielding the solution

$$I_j(t)=I_j(0),\quad \theta_j(t)=\omega_j(I_1(0),\dots,I_d(0))t+\theta_j(0), \tag{6.69}$$

giving the d periodic motions $\theta_j(t)$ on the torus defined by $(I_1(0),\dots,I_d(0))$, as depicted in Figure 6.8. It should be emphasized that even though such a coordinate transformation exists, it remains theoretical, and the value of the frequencies is available only in theory. In what follows we aim to identify exactly the frequency and the change of variables using deep learning.

Action-angle coordinates can be leveraged to provide a global linearization of an integrable Hamiltonian system. First, suppose we have an initial condition $(p(0),q(0))$ and the point $(p(h),q(h))$ is the solution to the initial value problem under the dynamics of (6.63) $h>0$ time units in the future. Then, if we define

$$\psi_j(p,q):=I_j(p,q)\mathrm{e}^{\mathrm{i}\theta_j(p,q)},\qquad j=1,\dots,d, \tag{6.70}$$

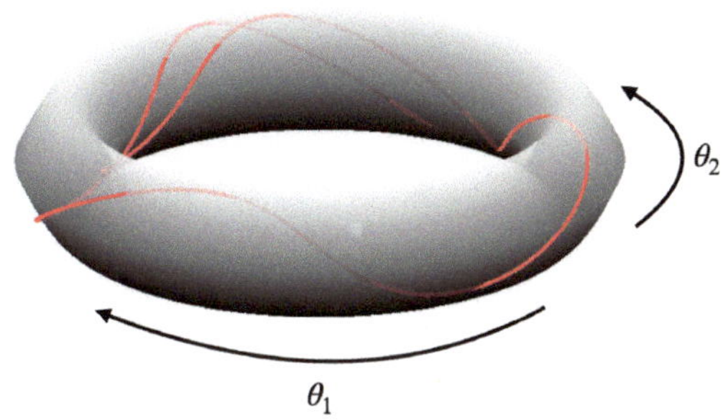

Figure 6.8. *Action-angle coordinates provide a nonlinear change of variables that describes motion on a torus. A sample trajectory is plotted in red.*

we find it satisfies $\psi_j(p(h), q(h)) = \mathrm{e}^{\mathrm{i}\omega_j(I(0))h}\psi_j(p(0), q(0))$. Indeed, this follows from the action-angle formulation (6.68) because

$$\begin{aligned}\psi_j(p(h), q(h)) &= I_j(h)\mathrm{e}^{\mathrm{i}\theta_j(h)} \\ &= I_j(0)\mathrm{e}^{\mathrm{i}\omega_j(I_1(0),\ldots,I_d(0))h+\mathrm{i}\theta_j(0)} \\ &= \mathrm{e}^{\mathrm{i}\omega_j(I_1(0),\ldots,I_d(0))h}\psi_j(p(0), q(0))\end{aligned} \tag{6.71}$$

since $I_j(h) = I_j(0)$ for all $h > 0$. Therefore, the $\psi_j : \mathbb{R}^{2d} \to \mathbb{C}$ can be interpreted as Koopman eigenfunctions of the flow h-map for the continuous-time Hamiltonian system (6.63). However, there is one subtlety in that the associated eigenvalue $\mathrm{e}^{\mathrm{i}\omega_j(I(0))h}$ depends on the trajectory, coming from the value of ω_j that is selected by the initial action values.

In order to implement the action-angle–based global linearization scheme using the functions (6.70) we must augment the autoencoder setup described at the beginning of this section. First, the transformation from the state-space coordinates (p, q) to the action-angle coordinates (I, θ) is a homeomorphism, but the inverse is in general nonlinear, thus requiring the decoder to be nonlinear as well. This is a trade-off that allows for the linearized state-space dimension, $2r$, to be kept equal to the original Hamiltonian differential equation state-space dimension, $2d$. Furthermore, from (6.71) we see that the eigenvalues $\mathrm{e}^{\mathrm{i}\omega_j h}$ of the observables ψ_j depend on the initial condition. Therefore, we augment the autoencoder neural network architecture to include another neural network, denoted $\Omega : \mathbb{R}^d \to \mathbb{R}^d$, trained so that

$$\Omega(y_1^2 + y_2^2, y_3^2 + y_4^2, \ldots, y_{2d-1}^2 + y_{2d}^2) = \begin{bmatrix} \omega_1(I_1, I_2, \ldots, I_d) \\ \vdots \\ \omega_d(I_1, I_2, \ldots, I_d) \end{bmatrix}. \tag{6.72}$$

Indeed, with $y_{2j-1} = \mathrm{Re}(\psi_j)$ and $y_{2j} = \mathrm{Im}(\psi_j)$ we get $I_j^2 = y_{2j-1}^2 + y_{2j}^2$ for all $j = 1, \ldots, d$, and we identify the phases from the values of the linearized coordinate variables $(y_1, y_2, \ldots, y_{2d-1}, y_{2d})$.

Example 6.8. To demonstrate the method of using an auxiliary neural network to identify the trajectory-dependent Koopman eigenvalues, we consider the Kepler problem. The Kepler problem is a particular case of the two-body problem where the bodies interact via a central force that varies as the inverse square of the distance between them. Such a problem is important to celestial mechanics since Newtonian gravity obeys an inverse square law, and therefore the dynamics of the Kepler problem can be used to describe a pair of binary stars orbiting each other, a satellite moving around a planet, or a planet moving around its sun.

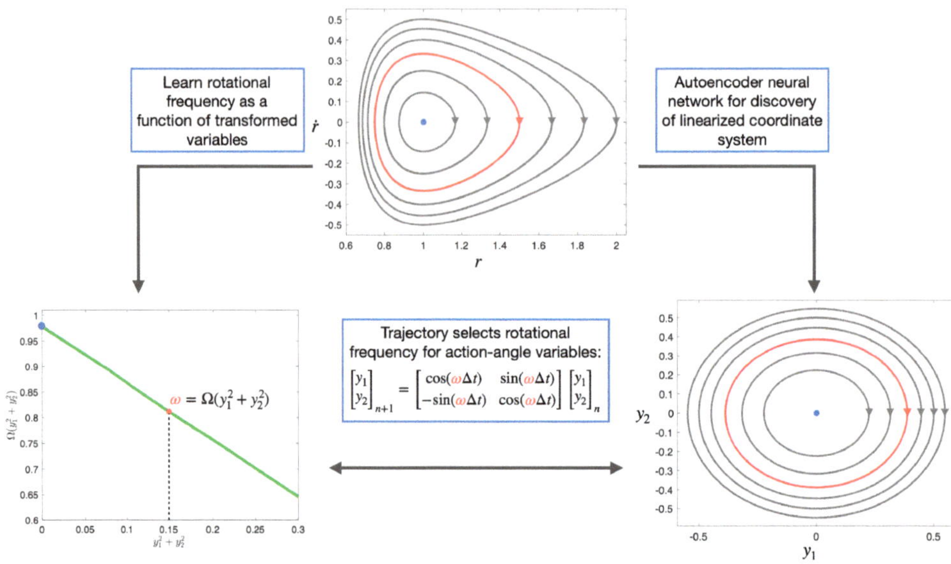

Figure 6.9. *Using autoencoders to discover a linearizing transformation of the Kepler problem. The network simultaneously discovers the rotational frequency of a periodic trajectory (bottom, left) along with a transformation to a linearized coordinate frame* (y_1, y_2) *based on action-angle coordinates of the Kepler problem (bottom, right). Top: A phase portrait of the Kepler problem* (6.73) *in the* $(r, \dot{r})$*-plane.*

After scaling out all constants, we may describe the time-dependent distance between the two bodies of the Kepler problem, denoted $r = r(t) \geq 0$, by the second-order ODE

$$\ddot{r} = \frac{1}{r^3} - \frac{1}{r^2}. \tag{6.73}$$

Equation (6.73) represents a Hamiltonian system with Hamiltonian function given by

$$\mathcal{H}(r, \dot{r}) = \frac{\dot{r}^2}{2} + \frac{1}{2r^2} - \frac{1}{r}. \tag{6.74}$$

Here $p = \dot{r}$ is the generalized momentum, and $q = r$ is the generalized position using (6.63). In the top panel of Figure 6.9 we provide a characteristic phase-plane portrait of the dynamics of (6.73) in the $(r, \dot{r})$-plane. There is a unique steady state $(r, \dot{r}) = (1, 0)$ representing the bodies orbiting each other at a constant distance. The remaining trajectories are periodic, physically representing the bodies orbiting each other with elliptic orbits, which is typical of binary stars.

Since $d = 1$ for the Kepler problem, the existence of the Hamiltonian structure is sufficient to deduce that it is an integrable system. Furthermore, the Hamiltonian level sets are all closed curves, leading to the periodic motion shown in Figure 6.9, thus meaning that there exists an action-angle formulation of the Kepler problem. We leverage this knowledge to employ a neural network to discover a linearizing coordinate transformation for the system.

Data is generated by integrating (6.73) from random initial conditions $(r(0), \dot{r}(0)) = (r_0, 0)$, with $r_0 \in [1, 2]$ and step size $h = 0.2$. Since the step from one time unit to the next in the numerical data is small, we employ a loss function that considers 30 steps into the future, i.e., $S = 30$ in the autoencoder losses (6.20), (6.21), and (6.22). This is to ensure that the loss properly encodes the timestepping of the flow h-map. As described above, we seek to identify an invertible transformation to the linear map

$$\begin{bmatrix} y_1 \\ y_2 \end{bmatrix}_{n+1} = \begin{bmatrix} \cos(\Omega(y_1^2+y_2^2)h) & \sin(\Omega(y_1^2+y_2^2)h) \\ -\sin(\Omega(y_1^2+y_2^2)h) & \cos(\Omega(y_1^2+y_2^2)h) \end{bmatrix} \begin{bmatrix} y_1 \\ y_2 \end{bmatrix}_n, \tag{6.75}$$

where $\Omega : \mathbb{R} \to \mathbb{R}$ is an auxiliary neural network trained in parallel to identify the trajectory-specific frequency of the periodic motion. Furthermore, linearizing (6.73) about $(1, 0)$ gives

$$\ddot{r} = -r, \tag{6.76}$$

which has eigenvalues $\pm$i. This means that periodic trajectories of (6.73) taken sufficiently close to the equilibrium will have frequency $\sim$1. Thus, we should expect $\Omega(r) \to 1$ as $r \to 0^+$.

Figure 6.9 presents results from a trained autoencoder that linearizes the dynamics of the differential equation (6.73). In particular, in the bottom left of Figure 6.9 we provide the learned frequencies of the periodic solutions of (6.73), identified from the auxiliary neural network Ω. For this trained network we have found $\Omega(0) =$ `0.97748`, which represents a 2.25% relative error from the true value of 1. The result of the trained network is that we can select an initial condition for (6.73), transform it using the encoder to the linearized coordinate variables (y_1, y_2), use these transformed values to identify the rotational frequency $\Omega(y_1^2 + y_2^2)$, iterate (6.75) for arbitrarily long to move forward in time, and finally transform each iterate back to the Kepler problem to produce the timestepped initial condition according to (6.73).

6.5 ▪ Discovering Conjugate Dynamics

Over the course of this book we have been introduced to various dictionary-based learning techniques. These include EDMD, SINDy, and PINNs. All of these methods require one to "guess" the right functions to work with when attempting to discover the underlying physics. Recall that the setting for these problems begins with data matrices

$$X = \begin{bmatrix} | & | & & | \\ X_1 & X_2 & \cdots & X_N \\ | & | & & | \end{bmatrix}, \quad Y = \begin{bmatrix} | & | & & | \\ Y_1 & Y_2 & \cdots & Y_N \\ | & | & & | \end{bmatrix} \in \mathbb{R}^{M\times N} \tag{6.77}$$

with the goal of identifying a function that maps data points $X_n \mapsto Y_n$ for each $n = 1, \dots, N$. As we did with SINDy in Section 3.1, we define a dictionary of $K \geq 1$ functions , $\mathcal{D} = \{\theta_1, \theta_2, \dots, \theta_K\}$, of mappings from $\mathbb{R}^M$ to $\mathbb{R}$. Then we could seek to identify coefficients

$$\Xi = \begin{bmatrix} | & | & & | \\ \Xi_1 & \Xi_2 & \cdots & \Xi_K \\ | & | & & | \end{bmatrix} \in \mathbb{R}^{M\times K} \tag{6.78}$$

to best approximate the dynamics $X_n \mapsto Y_n$ as the linear span of dictionary functions by

$$Y_n \approx \Xi_1\theta_1(X_n) + \Xi_2\theta_2(X_n) + \cdots + \Xi_K\theta_K(X_n). \tag{6.79}$$

It is important to note that one of the following two scenarios might arise:

1. One may be limited by a dictionary that includes too few functions (K is small) to accurately describe the dynamics.

2. One may use a dictionary so large ($K \gg 1$) that the discovered governing equations are too complex to work with beyond plugging them into a computer for simulation.

Both of these are very real possibilities that accompany dictionary-based methods and have the potential to pose the user with the problem of sacrificing accuracy for interpretability, as with (1), or sacrificing interpretability for accuracy, as with (2). For example, using monomials in the dictionary with large M, the number of monomials required to sufficiently approximate the dynamics is likely extremely large, leading to the situation described in point (2) above.

We can strive to strike a balance between accuracy and interpretability using autoencoders. To achieve this balance we will wrap the discovery process in an autoencoder and attempt to discover a conjugate system to the one that governs the dynamics $X_n \mapsto Y_n$. This amounts to a method of **indirect discovery** of the dynamics, as opposed a direct discovery of the action taking X_n to Y_n. In particular, we will train encoder and decoder functions $\eta : \mathbb{R}^M \to \mathbb{R}^r$ and $\rho : \mathbb{R}^r \to \mathbb{R}^M$, respectively, with $r \leq M$ and as small as possible. This acts to project the dynamics of the data onto a lower-dimensional manifold in $\mathbb{R}^r$, possibly eliminating redundancies in the data. Simultaneously, we represent the dynamics of the conjugate mapping g in the latent space $\mathbb{R}^r$ in the span of the dictionary $\tilde{\mathcal{D}} = \{\tilde{\theta}_1, \tilde{\theta}_2, \dots, \tilde{\theta}_K\}$ so that $\tilde{\theta}_k : \mathbb{R}^r \to \mathbb{R}$ for each $k = 1, \dots, K$. The coefficients $\Xi_k \in \mathbb{R}^r$ are trainable network parameters used to identify a conjugate map

$$g = \Xi_1\tilde{\theta}_1 + \Xi_2\tilde{\theta}_2 + \cdots + \Xi_K\tilde{\theta}_K. \tag{6.80}$$

The loss function takes the form of the standard autoencoder loss function, as in Section 6.2.

Notice that without including dimensionality reduction, i.e., $r = M$, this process of discovering conjugate dynamics should be expected to perform at least as well as a direct discovery method such as SINDy and likely better. The reason for this is that one could simply have both the encoder and decoder functions be the identity transformation, thus reducing discovery to the standard process of fitting the data to the span of library functions. The advantage now is that the encoder function $\eta : \mathbb{R}^M \to \mathbb{R}^r$ can deform and reshape the data in such a way that they best fit the dictionary elements supplied for the discovery process. For example, imagine that the data supplied is gathered from the sine map (6.26) of Example 6.4. Using a dictionary of monomials could yield a truncated Taylor expansion of the sine function, as in

$$\sin(\pi x) \approx \pi x - \frac{\pi^3}{6}x^3 + \frac{\pi^5}{120}x^5 - \dots, \tag{6.81}$$

while the encoder function can learn the optimal conjugacy between the sine function and the logistic function (6.6) so that discovery of the map g only requires monomials up to degree 2. This greatly reduces the degrees of freedom endowed to the system by the size of the dictionary while also producing a simple and interpretable system based on the dictionary elements.

Example 6.9. In this example we consider the Rössler system

$$\begin{aligned}\dot{x}_1 &= -x_2 - x_3,\\ \dot{x}_2 &= x_1 + 0.1x_2,\\ \dot{x}_3 &= 0.1 + x_3(x_1 - \mu),\end{aligned} \tag{6.82}$$

where $\mu \in \mathbb{R}$ is a bifurcation parameter. System (6.82) has the property that when x_1 crosses zero on the attractor, x_3 numerically appears to as well. Therefore, we may introduce a Poincaré section at $x_1 = 0$ with the crossing condition $\dot{x}_1 > 0$ so that the related Poincaré map is only a function of the scalar x_2 variable. Figure 6.10 provides plots of the Poincaré maps over differing μ using data gathered from numerical integration of the ODE. Importantly, one notices that the data appears to lie along a unimodal map, much like those explored in the examples of Section 6.2. Therefore, informed by those examples and analytical results establishing conjugacies between unimodal maps and the logistic map [108, 197], we attempt to learn the dynamics on the Poincaré section of the Rössler system using a dictionary up to only quadratic monomials.

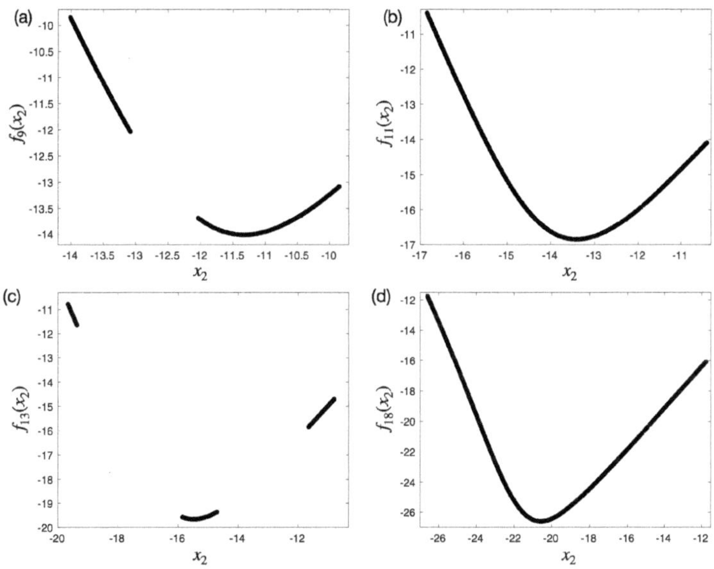

Figure 6.10. *Poincaré section data of the Rössler system* (6.82) *for* (a) $\mu = 9$, (b) $\mu = 11$, (c) $\mu = 13$, *and* (d) $\mu = 18$. *Plotted is data from the Poincaré map* $f_\mu(x_2)$, *i.e., the next iterate in the section, against the current iterate* x_2.

Table 6.2. *Comparison of discovered Rössler Poincaré maps using an autoencoder versus fitting the* x_2 *data directly to a quadratic function. Results are presented over a range of chaotic parameter values* μ *for illustration. Also provided are the corresponding values of* ξ_1 *in the logistic map* (6.84) *that both discovered maps are conjugate to.*

	Autoencoder		Quadratic fit	
μ	Mapping	ξ_1-value	Mapping	ξ_1-value
9	$3.6075y - 4.9044y^2$	3.6075	$1.0219 - 2.9885x_2 + 2.2063x_2^2$	3.6219
11	$3.8302y - 4.4801y^2$	3.8302	$1.0217 - 3.4481x_2 + 2.9242x_2^2$	3.7985
13	$3.8502y - 6.5346y^2$	3.8502	$1.0075 - 3.5638x_2 + 3.1613x_2^2$	3.8456
18	$3.9661y - 4.6718y^2$	3.9661	$0.9776 - 3.5219x_2 + 3.4089x_2^2$	3.6737

To learn a quadratic representation of the Rössler Poincaré map dynamics, we use an autoencoder neural network to identify a conjugacy with a function

$$g(y) = \xi_1 y + \xi_2 y^2, \tag{6.83}$$

with parameters $\xi_1, \xi_2 \in \mathbb{R}$ to be learned by the network during training. A simple linear transformation establishes a conjugacy between (6.83) and the map

$$\tilde{g}(y) = \xi_1 y\left(1 + \frac{\xi_2}{|\xi_2|}y\right), \tag{6.84}$$

which is exactly the logistic map when $\xi_2 < 0 < \xi_1$, with the leading parameter value ξ_1 determining the dynamics. The map (6.83) simply provides the neural network with an extra degree of freedom that provides more opportunity to establish a conjugacy beyond the restricted logistic setting of $\xi_2 = -\xi_1$.

We present the results of our data-driven discovery using autoencoders in Table 6.2 using the chaotic parameter values $\mu = 9, 11, 13,$ and 18. For comparison, we also provide the result

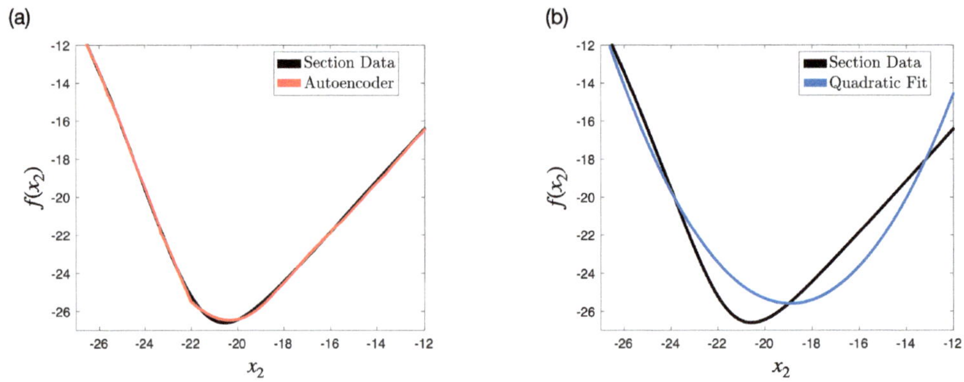

Figure 6.11. *Fitting the Poincaré section data with an autoencoder versus a quadratic curve with* $\mu = 18$. (a) *The Poincaré map data (black) against the output of the autoencoder* $f(x_2) \approx \rho \circ g \circ \eta(x_2)$, *with* $g(y) = 3.9661y - 4.6718y^2$ *the discovered conjugate map.* (b) *Fitting the Poincaré section data to a quadratic curve gives* $0.9776 - 3.5219x_2 + 3.4089x_2^2$, *which is a poor approximation of the true Poincaré map.*

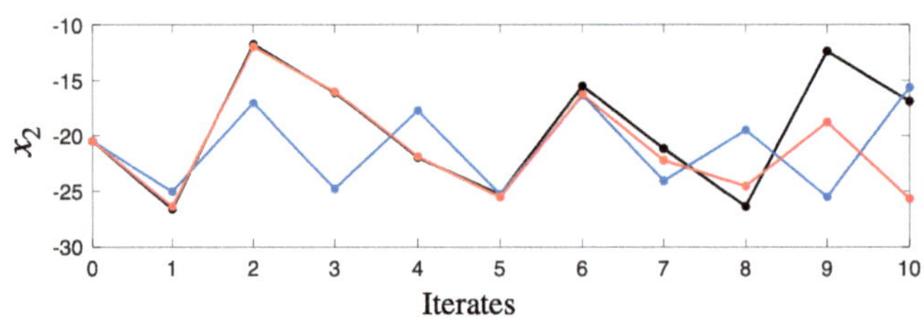

Figure 6.12. *Forecasting the Rössler Poincaré section data (black) at* $\mu = 18$ *using the autoencoder neural network (red) and the fitted quadratic curve (blue).*

of fitting the data to a quadratic curve, which amounts to a SINDy representation of the section data using the quadratic library $\{1, x_2, x_2^2\}$. The resulting SINDy maps can be transformed to a logistic function of the form (6.84) by a linear change of variable, and so the resulting values of ξ_1 are presented to compare the results. The values of ξ_1 between the conjugate and SINDy maps show the largest difference for $\mu = 11$ and $\mu = 18$. Figure 6.10 shows that these parameter values have the fullest chaotic attractors, and so fitting a quadratic curve to the data would only succeed if the curve were indeed truly quadratic. Figure 6.11(b) demonstrates the poor performance of this curve fitting, while panel (a) shows the fitting of the map data with the autoencoder conjugacy $\rho \circ g \circ \eta$, with $g(y) = 3.9661y - 4.6718y^2$. The performance of the autoencoder is greatly improved, although there is some imperfection near the global minimum of the data.

We can also forecast the data using the discovered conjugate mapping. In Figure 6.12 we provide the autoencoder forecast of the Poincaré mapping data at $\mu = 18$. The forecast is obtained by iterating forward in the conjugate map, meaning that the nth forward iterate is obtained by $\rho \circ g^n \circ \eta$. This is done to demonstrate that the dynamics of the system are truly encoded in the conjugate map g, while the encoder and decoder are simply for transforming back and forth between the section data and the map g. For comparison, we also provide the forecast using the fitted quadratic curve, which from Figure 6.11(b) we expect to perform poorly. The autoencoder forecast remains accurate for at least 6 full iterates in the section, eventually falling off. This falloff should be expected, as the Poincaré map is chaotic and therefore sensitive to perturbation and approximation. Nonetheless, a prediction horizon of 6 iterates of the Poincaré map represents a prediction of approximately 42 time units into the future of the continuous-time system (6.82).

The previous example highlights how autoencoders can transform data to fit a limited, or "simple," dictionary. Another utility of the autoencoder discovery framework is to reduce the dimension of the data. Recall that in Section 6.4 we wanted to inflate the dimension of the data in order to lift the dynamics to a linear coordinate frame. Here our goal is to produce nonlinear models of the data but to do this with as few degrees of freedom as possible. An important example of why one would like to do this comes from [54], where dimensionality reduction is used to reduce simulated video of a pendulum swinging to the two-dimensional ODE that describes its motion. Similarly, [38] uses this dimensionality reduction to relate the high-dimensional Kuramoto–Sivashinsky equation to the dynamics of a one-dimensional logistic map. This relationship between high-dimensional complex systems and the logistic map is further explored in [161], which proposes the logistic map as a normal form for damped-driven systems.

Identifying candidates for the latent dimension r can be a nontrivial task, and in many cases one is required to treat r as a hyperparameter and optimize fitting over it. However, in the case of chaotic Poincaré section data we can rely on analysis that relates the unstable directions of the attractor to its dimension through the **Lyapunov exponents**. To do this we first consider a continuous-time dynamical system

$$\dot{x} = F(x) \tag{6.85}$$

and a solution $x(t)$. Lyapunov exponents describe the rates of expansion and contraction in the transverse dimensions along the trajectory $x(t)$ using the Jacobian $J(t) := F_x(x(t))$. The tangent vectors evolve in time according to the linear ODE

$$\dot{M}(t) = J(t)M(t), \tag{6.86}$$

with initial condition $M(0) = I$, the identity matrix, while $M(t)$ is the same dimension as $J(t)$. The Lyapunov exponents are then defined to be the eigenvalues of the matrix

$$\Lambda := \lim_{t\to\infty} \log(M(t)M(t)^\intercal), \tag{6.87}$$

which are always guaranteed to exist by Oseledet's theorem [209]. Moreover, the matrix Λ is symmetric, meaning that the Lyapunov exponents are always real. Positive exponents correspond to dimensions of exponential separation of trajectories starting nearby $x(t)$, while negative exponents correspond to dimensions of exponential contraction toward $x(t)$.

For steady-states the Lyapunov exponents are just the real part of the eigenvalues of their Jacobian matrix, and when $x(t)$ is periodic, the exponents are the real parts of the Floquet exponents. However, Lyapunov exponents are most typically used for chaotic systems. An important part of Oseledet's theorem is that the Lyapunov exponents are independent of the trajectory $x(t)$ when the system is chaotic. Moreover, the presence of at least one positive Lyapunov exponent is a common indicator that the system is chaotic, as it says that all nearby trajectories separate at an exponential rate in at least one dimension. This separation is more commonly referred to as **sensitive dependence on initial conditions** and is one of the defining features of chaos.

Coming back to identifying the latent dimension $r \geq 1$ for a chaotic Poincaré map, we begin by ordering the Lyapunov exponents from largest to smallest,

$$\lambda_1 \geq \lambda_2 \geq \cdots \geq \lambda_d, \tag{6.88}$$

where $d \geq 1$ indicates the dimension of the state space of the differential equation (4.1). The **Kaplan–Yorke dimension** of a chaotic system is given by

$$D_{\text{KY}} = k + \sum_{j=1}^{k} \frac{\lambda_j}{|\lambda_{k+1}|}, \quad \text{with} \quad \begin{cases} \lambda_1 + \cdots + \lambda_k \geq 0, \\ \lambda_1 + \cdots + \lambda_k + \lambda_{k+1} < 0. \end{cases} \tag{6.89}$$

The Kaplan–Yorke conjecture [138, 92] states that the dimension of a chaotic attractor (in continuous time) is equal to its Kaplan–Yorke dimension. Since we are interested in Poincaré mappings here, we can estimate the latent dimension phase space to be

$$r \approx D_{\mathrm{KY}} - 1, \tag{6.90}$$

where we subtract 1 because the phase space of the Poincaré map is a $(d-1)$-dimensional hyperplane. The approximation comes from the fact that D_{KY} is rarely an integer, resulting in noninteger attractor dimensions. Let us now employ these ideas with the following example.

Example 6.10. To examine the capability for autoencoders to simultaneously perform dimensionality reduction, variable conjugation, and system identification, we examine the system

$$\begin{aligned}\dot{x}_1 &= \mu x_1 - x_2 x_3,\\ \dot{x}_2 &= -\nu x_2 + x_1 x_3,\\ \dot{x}_3 &= \Gamma - x_3 + x_1 x_3,\end{aligned} \tag{6.91}$$

which was introduced by Gissinger [96]. The Poincaré section originally used by Gissinger to analyze the dynamics of (6.91) is slightly tedious, and therefore we will rescale the variables to simplify the presentation. To this end, we apply the linear change of variable

$$\begin{aligned}x_1 &= \hat{x}_1\sqrt{\nu + \Gamma\sqrt{\frac{\nu}{\mu}}},\\ x_2 &= \hat{x}_2\sqrt{\mu + \Gamma\sqrt{\frac{\mu}{\nu}}},\\ x_3 &= \Gamma - \hat{x}_3(\sqrt{\mu\nu} + \Gamma).\end{aligned} \tag{6.92}$$

In the hatted variables Gissinger's Poincaré section becomes $\hat{x}_1 + \hat{x}_2 = 0$ with the crossing condition $\dot{\hat{x}}_1 + \dot{\hat{x}}_2 > 0$. The result of this linear change of variable is to shift one equilibrium of (6.91) to the origin, while the two nontrivial equilibria are given by $(\hat{x}_1, \hat{x}_2, \hat{x}_3) = (\pm 1, \mp 1, 1)$.

Throughout we fix the parameter values $(\mu, \nu, \Gamma) = (0.12, 0.1, 0.85)$, resulting in a chaotic attractor for system (6.91). At these parameter values the Lyapunov exponents are given to two decimal places by $\lambda_1 = 0.07$, $\lambda_2 = 0.00$, and $\lambda_3 = -1.05$, which can be computed by following [287]. The presence of $\lambda_1 > 0$ indicates that the system is chaotic, while the Lyapunov exponent at 0 is always present and corresponds to transverse translations along the trajectory in time. This gives a Kaplan–Yorke dimension of 2.07, for which we will take the latent dimension to be $r = 1$. Although we will inevitably lose some information coming from the noninteger part of the Kaplan–Yorke dimension, Figure 6.13 demonstrates that the Poincaré mapping $f : \mathbb{R}^2 \to \mathbb{R}^2$ nearly splits as $f(x_2, x_3) = [f_1(x_2), f_2(x_3)]^\intercal$ with f_1 and f_2 similar scalar mappings.

Guided by Figure 6.13, we use an autoencoder neural network to identify a conjugacy with the Poincaré mapping dynamics of (6.91) with a cubic map. The resulting discovery process learns a conjugacy of the two-dimensional Gissinger section data with the one-dimensional map

$$g(y) = 8.53442y - 18.29989y^2 + 9.81721y^3, \tag{6.93}$$

which is plotted in Figure 6.14. We can see that the map (6.93) takes the interval $0 \le y \le 1.17952$ back into itself. To better understand its dynamics, we can subdivide the domain into three subintervals I_0, I_1, and I_2, broken up at $y = 0.31103$, the location of the local maximum, and $y = 0.93168$, the location of the local minimum. This subdivision is shown in Figure 6.14.

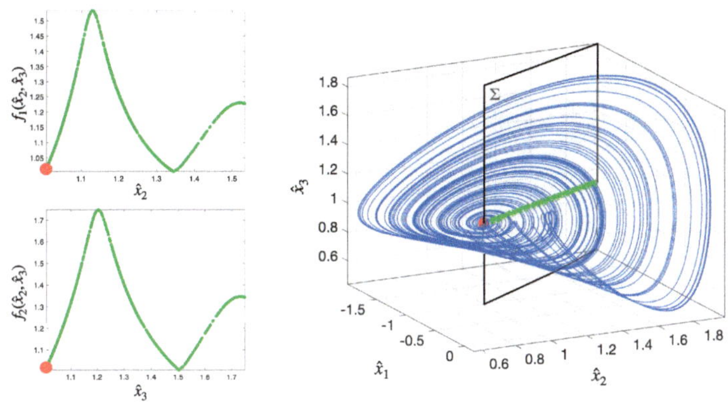

Figure 6.13. *The Gissinger system* (6.91) *has nearly one-dimensional Poincaré section dynamics. Left: The Poincaré mapping components $f = (f_1, f_2)$ as a function of $(\hat{x}_2, \hat{x}_3)$ in the Poincaré section $\hat{x}_1 + \hat{x}_2 = 0$, with $\dot{\hat{x}}_1 + \dot{\hat{x}}_2 > 0$. Right: The chaotic attractor (blue) in the full three-dimensional phase space $(\hat{x}_1, \hat{x}_2, \hat{x}_3)$ with the Poincaré section indicated as Σ. The equilibrium $(-1, 1, 1)$ at the center of the attractor is denoted by a red dot in all figures.*

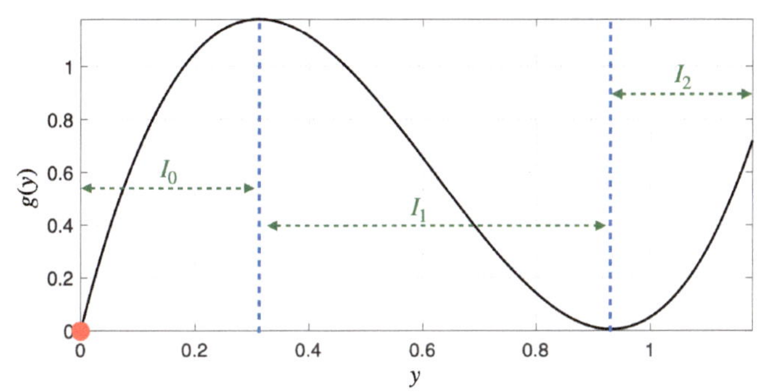

Figure 6.14. *The cubic map* (6.93) *identified as (approximately) conjugate to the dynamics in the Poincaré section of the Gissinger system* (6.91)*. The dynamics of the map are best understood by partitioning the domain into three distinct intervals, I_0, I_1, and I_2, broken up by the critical points of the function g. The red dot indicates the fixed point $y = 0$ of the map that corresponds (under conjugation) to the equilibrium $(\hat{x}_1, \hat{x}_2, \hat{x}_3) = (-1, 1, 1)$ indicated by a red dot in Figure* 6.13.

The function g maps I_0 and I_1 to the entire interval, while $g(1.17952) = 0.71679$, meaning that

$$I_0 \subsetneq g(I_2) \subsetneq I_0 \cup I_1. \tag{6.94}$$

Thus, the interval $0 \leq y \leq 0.71679$ has exactly three preimages under g, while the interval $0.71679 < y \leq 1.17952$ has only two.

The division of the domain of the conjugate map (6.93) is useful for identifying and classifying the periodic orbits of the Gissinger system (6.91). Indeed, if $\{y_1, y_2, \ldots, y_p\}$ is a p-cycle of g, then we can use the decoder ρ to bring these values back to the original two-dimensional Poincaré section. The points $\{\rho(y_1), \rho(y_2), \ldots, \rho(y_p)\}$ are approximately the locations of an unstable periodic orbit of the Gissinger ODE intersected with the Poincaré section. Periodic orbits intersect the Poincaré section at $p \geq 1$ unique points before completing one full period, and these intersections can be used to identify the full continuous-time orbit in a process that is explained at the end of this section. Figure 6.15 provides all identified periodic orbits up to four

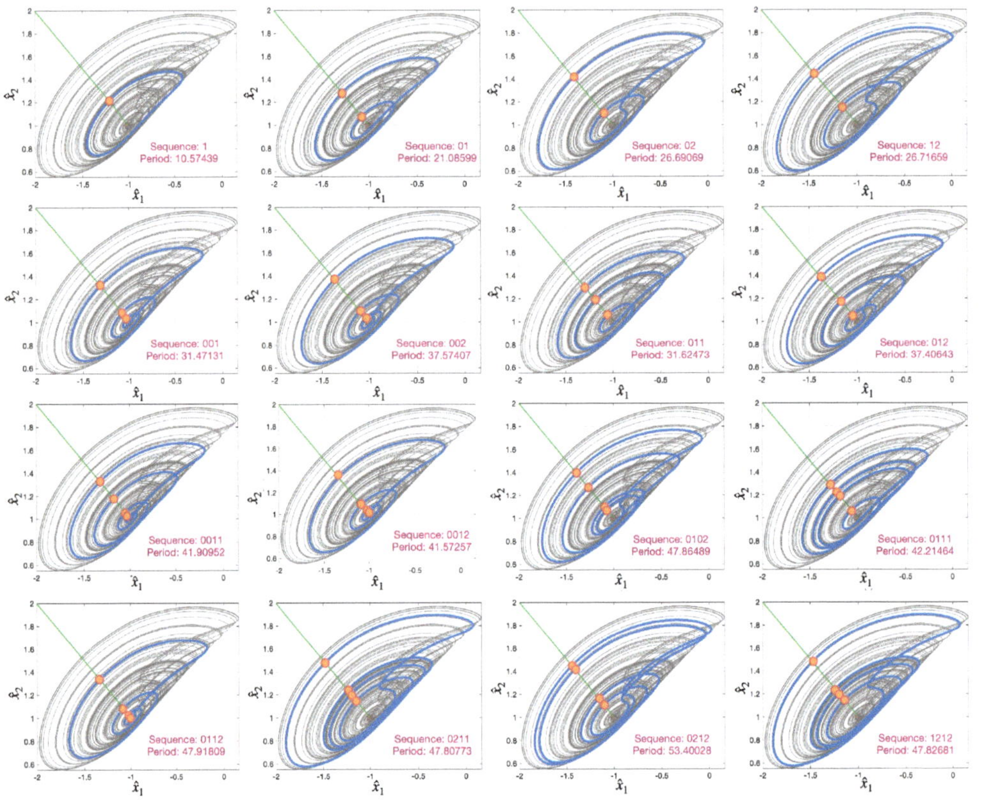

Figure 6.15. *Unstable periodic orbits of the chaotic Gissinger system* (6.91) *and their temporal periods. All periodic orbits are plotted in blue with the chaotic attractor in gray. The green line represents the Poincaré section projected onto the* $(\hat{x}_2, \hat{x}_3)$*-plane, and orbits are identified by their symbolic sequence, whose length corresponds with the number of times they intersect the Poincaré section. Intersections are denoted by orange circles.*

intersections with the Poincaré section. Orbits are uniquely identified by a symbolic sequence $a_1 a_2 \cdots a_p$ with $a_j \in \{0, 1, 2\}$ so that

$$a_j = \begin{cases} 0, & y_j \in I_0, \\ 1, & y_j \in I_1, \\ 2, & y_j \in I_2. \end{cases} \tag{6.95}$$

The sequence corresponding to each periodic orbit is provided in Figure 6.15, along with their period. Sequences with more 2's have longer period, coming from the observation that 2's correspond to large deviations of the periodic orbit from the equilibrium $(\hat{x}_1, \hat{x}_2, \hat{x}_3) = (-1, 1, 0)$.

One may continue to identify periodic orbits with longer and longer sequences which can in turn be used to implement the *periodic orbit theory* of Cvitanović and collaborators [67]. This theory allows one to assemble measurable global averages over the chaotic attractor as weighted sums of the periodic orbits and determine the natural measure to give the recurrence statistics along the attractor in phase space. Thus, we may use the unstable periodic orbits of a chaotic system to answer important questions from ergodic theory. Furthermore, the work [292] uses the unstable periodic orbits in a turbulent fluid model to reduce a PDE to a 17-node Markov model representing jumping from one neighborhood of a periodic orbit to another.

In the previous example we used the Lyapunov exponents to target a reduced-order scalar mapping to approximate the dynamics and topology of a chaotic system. Scalar mappings are particularly well studied and come with tools such as symbolic dynamics and kneading theory that help us to fully understand their dynamics [12, 73, 197, 228]. In situations when there are two or more positive Lyapunov exponents, so-called hyperchaos, the situation becomes more complex. This is because the Kaplan–Yorke dimension is at least 3, requiring a latent dimension of at least 2. It can then be difficult to determine an appropriate library to build the conjugate mapping. Furthermore, multivariable mappings are not as well-understood as their scalar counterparts, meaning that the analysis of the learned conjugate mapping may require some creativity.

6.6 ▪ Optimal Discovery with Autoencoders

Autoencoders have the potential to significantly alter our understanding of nonlinear dynamics. As we saw in this chapter, autoencoders can leverage the extensive work on analyzing dynamical systems to produce methods that complement, not replace, the study of time-dependent systems. They allow one to develop innovative approaches that can bridge the gap between specific systems and well-understood prototypical models. This might be as simple as the explicit identification of optimal changes of variable that recast difficult problems in simpler forms. This is the case of conjugacies with the logistic map, linearization techniques like the Hartman–Grobman theorem and action-angle variables, as well as normal forms. However, autoencoders can also be combined with system identification, which we saw can include discovering high-dimensional linear dynamics using Koopman observables or low-dimensional nonlinear models.

The uses of autoencoders for dynamic data analysis go far beyond the applications presented in this chapter. For example, [166] employs autoencoders to disentangle periodic signals for long-range forecasting. This work uses variable transformations that are inspired by Koopman operator theory to provide an optimal basis representation to provide a sparse Fourier expansion of the data. The work [23] combines delay coordinates (see Section 2.3) with autoencoders to implement Takens's delay embedding theorem to reconstruct attractors from scalar data, while [63] combines autoencoders and SINDy to discover parameter-dependent models from data. The works [175, 176] contain another notable application of autoencoders to dynamical systems; in this application the autoencoders learn low-dimensional ODEs on inertial manifolds to forecast discretizations of PDEs.

The future of autoencoders applied to dynamical systems looks bright. More and more abstract mathematical concepts are becoming accessible through this deep learning architecture. This is most apparent through new methods that seek to learn local charts of manifolds embedded in high-dimensional space [251, 252]. In these applications one uses local charts to construct a patchwork of the global manifold structure using a partition of unity. This provides dimensionality reduction of observed data to lower-dimensional spaces with nontrivial topology. This is particularly exciting for dynamical systems, as the reduction to lower-dimensional latent manifolds can be paired with model discovery to learn the intrinsic dynamics of a high-dimensional system [89]. This may lead to applications on turbulent fluid models with low-dimensional chaotic attractors embedded in their phase space. Autoencoders therefore have the capability to identify nontrivial topologies of attractors along with the dynamics on them.

6.7 ▪ Appendix: Root Finding for Periodic Orbits

Periodic solutions of (4.1) satisfy $x(t+T) = x(t)$ for some $T > 0$. In practice the period T is rarely known beforehand, and so both the period and the orbit need to be identified together. To do this we rescale the independent variable as $t = sT$ so that the ODE (4.1) becomes

$$\frac{\mathrm{d}x(s)}{\mathrm{d}s} = TF(x(s)). \tag{6.96}$$

The period T becomes a parameter, while the periodic criterion $x(t+T) = x(t)$ for (4.1) is equivalent to a solution of (6.96) satisfying $x(s+1) = x(s)$ for all s. Thus, the period is fixed to be exactly 1, and we are required to find a T at which (6.96) has a 1-periodic solution.

Periodic orbits of (6.96) can be equivalently viewed as roots of the system of equations [80]

$$\begin{aligned} \frac{\mathrm{d}x(s)}{\mathrm{d}s} - TF(x(s)) &= 0, \\ x(1) - x(0) &= 0, \\ \int_0^1 \left\langle \frac{\mathrm{d}x_0(s)}{\mathrm{d}s}, x_0(s) - x(s) \right\rangle \mathrm{d}s &= 0. \end{aligned} \tag{6.97}$$

Here $x_0(s)$ corresponds to an initial guess at the solution and is used in the final equation to fix a frame of reference that removes the degeneracy coming from the translational symmetry in s. That is, if $x(s)$ is a periodic solution of (6.96), then so is $x(s+\tau)$ for any $\tau \in \mathbb{R}$. The final equation in (6.97) fixes a value of τ to identify the periodic solution uniquely. The first equation guarantees the trajectory solves the given ODE, and the second equation guarantees 1-periodicity in s. The problem (6.97) can be discretized in s to get a system of nonlinear equations for which any number of root-finding algorithms can be used to identify a solution $(x(s), T)$ numerically.

Implementing a root-finding procedure to numerically identify solutions of (6.97) requires an initial guess, $x_0(s)$. Moreover, the ability to converge to a solution requires an initial guess that is sufficiently close to the desired solution. This is where the identified conjugate mapping $g(y)$ can help. Suppose that $\{y_1, y_2, \dots, y_p\}$ is a p-cycle of the map g so that $\{\rho(y_1), \rho(y_2), \dots, \rho(y_p)\}$ approximately represents the locations of the $p \geq 1$ intersections of a periodic orbit with a Poincaré section. For each $j = 1, \dots, p$ we may consider the initial value problem

$$\begin{aligned} \dot{x} &= F(x), \\ x(0) &= \rho(y_j), \end{aligned} \tag{6.98}$$

which leads to a trajectory that starts in the Poincaré section and remains close to the true periodic orbit for some finite time horizon that depends on the Lyapunov exponents. We flow each solution of (6.98) forward in t until it reaches the Poincaré section again. Since these trajectories started close to a periodic orbit, they should flow from $\rho(y_j)$ approximately to $\rho(y_{j+1})$, with indices taken modulo p. Thus, concatenating all p of these trajectories that flow from the Poincaré section back to it provides an initial guess for the periodic orbit that is stitched together with jump discontinuities in the Poincaré section at the concatenation points. This is exactly what was done to identify the unstable periodic orbits in Figure 6.15 for the Gissinger system.

Appendix A

Accompanying Scripts and Notebooks

Each chapter of this book has a number of MATLAB scripts and/or Jupyter notebooks that accompany it. These can be used to reproduce and extend the numerical results in each of the examples. The code is organized into folders for each chapter in the following repository:

https://bookstore.siam.org/ot201/bonus

Folders associated to Chapters 5 and 6 further contain trained neural networks that recreate the results in each section. The notebooks associated to each section contain blocks that can be used to load and check these results for the convenience of the user. However, one is always encouraged to train their own networks to measure performance and variation of outcomes against those presented here.

Below we list the scripts and notebooks associated to each section of the book, headed by the name of the folder the scripts can be found in. For sections with more than one associated script or notebook they are listed in order of appearance in that section. The contents of each folder in the associated repository are as follows:

Chapter 1: Dynamical Systems: Old and New
Section 1.3: `spiral_POD.m`

Chapter 2: Linear Evolution Models
Section 2.1: `DMD_Schrodinger.m`
Section 2.2: `windowDMD.m`
Section 2.3: `DelayDMD.m`
Section 2.4: `EDMD.m`
Section 2.5: `Kernel_DMD.m`

Chapter 3: Identifying Nonlinear Dynamics
Section 3.1: `SINDy.m`
Section 3.2: `SINDy_map.m`
Section 3.3: `deJong_control.m`, `Sprott_control.m`,
Section 3.4: `averaging.m`
Section 3.5: `conserved_quantities.m`

Chapter 4: Data-Driven Polynomial Optimization
Section 4.2: `MG_LyapFn.m`, `heteroclinic_UpperBnd.m`, `heteroclinic_LowerBnd.m`
Section 4.3: `Disc_LyapFn_Data.m`, `logistic_bounds.m`

Section 4.4: `pend_control.m`
Section 4.5: `invariant_measure.m`

Chapter 5: Learning Dynamics with Neural Networks

Section 5.2: `neural_network.m`
Section 5.3: `Forecast.ipynb`
Section 5.4: `Diffusion_PINN.ipynb`, `Bistable_PINN.ipynb`
Section 5.5: `Diffusion_Discovery.ipynb`

Chapter 6: Autoencoder Neural Networks

Section 6.2: `Tent2Logistic.ipynb`, `Tent2Sine.ipynb`,
Section 6.3: `NormalForm_sn.ipynb`, `NormalForm_pd.ipynb`
Section 6.4: `Global_Linearization.ipynb`, `ActionAngle.ipynb`
Section 6.5: `Rossler_conj.ipynb`, `Gissinger_conj.ipynb`

Appendix B
Singular Value Decomposition

The singular value decomposition (SVD) is a powerful method of matrix factorization that can be used throughout computational mathematics. The SVD can be thought of as a generalization of the method of diagonalizing a square matrix and applies to not only nondiagonalizable square matrices but also matrices of any size. Precisely, the SVD of a matrix $A \in \mathbb{C}^{M\times N}$ results in the factorization $A = U\Sigma V^*$, where $U \in \mathbb{C}^{M\times M}$ and $V \in \mathbb{C}^{N\times N}$ are unitary matrices, i.e., $U^{-1} = U^*$ and $V^{-1} = V^*$, and Σ is a rectangular diagonal matrix. See Figure B.1 for an illustration. We highlight the following properties of the SVD:

1. For any value of $M, N \geq 1$, every matrix $A \in \mathbb{C}^{M\times N}$ has an SVD.
2. The diagonal entries of Σ, denoted $\sigma_i := \Sigma_{i,i}$, are uniquely determined by A and are always nonnegative real numbers.
3. The singular values are always ordered from largest to smallest along the diagonal of Σ, meaning $\sigma_1 \geq \sigma_2 \geq \sigma_3 \ldots$.
4. The number of nonzero singular values is the rank of A.
5. If $r = \mathrm{rank}(A)$, then the first r columns of U form a basis for the range of A, and the last $N - r$ columns of V form a basis for the kernel.

It is worth noting that although the singular values are unique, the SVD of a matrix need not be. A simple way to see this is just to let $\tilde{U} = e^{i\theta}U$ and $\tilde{V} = e^{i\theta}V$ for any $\theta \in \mathbb{R}$ to find that again $A = \tilde{U}\Sigma\tilde{V}$, although more exotic nonuniqueness scenarios can be observed in degenerate cases.

B.1 ▪ Geometric Interpretation

Multiplication of a vector x by a matrix A is a linear transformation from one vector space to another, and thus we seek to understand the geometry of the transformation in terms of the SVD matrices U, Σ , and V. We first motivate this geometric interpretation with that of matrix diagonalization.

Recall that if A is a square, diagonalizable matrix, we can write it as $A = PDP^{-1}$, where D is a diagonal matrix. The columns of P span invariant subspaces for which the action of A on them is to simply multiply by a constant (the entries in D). Thus, for a vector x we have that $Ax = PDP^{-1}x$ acts by first applying $P^{-1}x$, which projects x onto each invariant subspace, then $D(P^{-1}x)$ simply stretches, compresses, or flips vectors in these subspaces, while the final

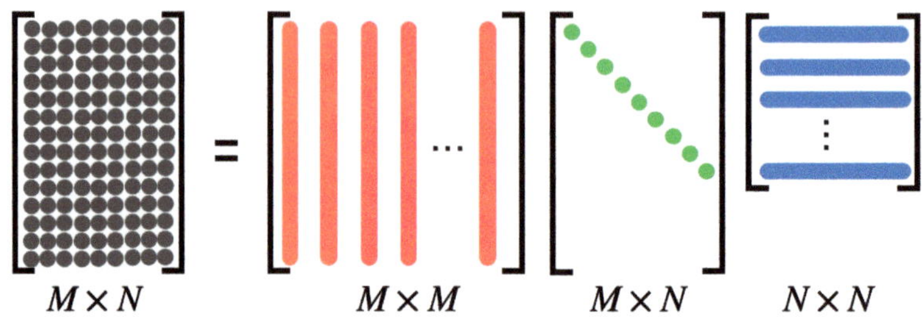

Figure B.1. *A visual illustration of the matrix factorization using the SVD. Matrix sizes are written underneath the illustrations.*

multiplication $P(DP^{-1}x)$ brings one back to the coordinate system in which x was originally posed.

Now let A be a general matrix, and let $A = U\Sigma V^*$ be its SVD. Since U and V are orthogonal matrices, their columns provide orthonormal bases for $\mathbb{C}^M$ and $\mathbb{C}^N$, respectively. Letting u_j and v_j be the jth columns of U and V, respectively, we have

$$A = U\Sigma V^* \implies AV = U\Sigma \implies Av_j = \sigma_j u_j \tag{B.1}$$

for each $j = 1, \ldots, \min\{M, N\}$ and $Av_j = 0$ for $j > \min\{M, N\}$. Thus, the decomposition $AV = U\Sigma$ demonstrates that one can find orthogonal basis vectors so that the jth basis vector of $\mathbb{C}^M$ is mapped into the jth basis vector of $\mathbb{C}^N$. Thus, similar to matrix diagonalization, with respect to these basis vectors the transformation $x \mapsto Ax$ acts as a diagonal matrix.

B.2 ▪ Compact SVD

In practice it is often the case that the matrix A has low rank, and so it can be advantageous to reduce the size of the matrix Σ to eliminate the abundance of zero singular values. This leads to what is sometimes referred to as the compact SVD, which provides the factorization $A = U_r\Sigma_r V_r^*$, where $U_r \in \mathbb{C}^{M\times r}$, $V_r \in \mathbb{C}^{N\times r}$, and $\Sigma_r \in \mathbb{R}^{r\times r}$ is a diagonal matrix of the $r = \text{rank}(A)$ nonzero singular values of A. The matrices U_r and V_r are now semiunitary in that $U_r^* U_r = V_r^* V_r = I_r$, the $r \times r$ identity matrix. Such matrices are formed from the SVD by eliminating the last $M - r$ and $N - r$ columns of U and V, respectively, and keeping only the first r rows and columns of Σ. Similarly, one can reduce the sizes of the SVD matrices to any s between $\text{rank}(A)$ and $\min\{M, N\}$ by replacing r with s in the above.

B.3 ▪ Low-Rank Approximations

The SVD provides a method of approximating a matrix with one of lower complexity, as measured by the rank. Precisely, in many practical applications (including dynamic mode decomposition) one wishes to solve the minimization problem

$$\underset{\hat{A}\in\mathbb{C}^{M\times N}}{\arg\min} \|A - \hat{A}\|_F \text{ such that } \text{rank}(\hat{A}) \leq s \tag{B.2}$$

for a given matrix A, where $\|\cdot\|_F$ denotes the Frobenius norm. The Eckart–Young–Mirsky theorem [78, 198] gives that the minimizer $\hat{A}$ is given by

$$\hat{A} = U_s\Sigma_s V_s^*, \tag{B.3}$$

where again the subscript denotes that we are only taking the first s columns of U and V and the first s rows and columns of Σ from the SVD $A = U\Sigma V$. Moreover, the error between these two matrices is given by

$$\|A - U_s\Sigma_s V_s^*\|_F = \sqrt{\sigma_{s+1}^2 + \cdots + \sigma_r^2}, \tag{B.4}$$

where $r = \operatorname{rank}(A)$. Mirsky also generalized this result to hold for any unitarily invariant matrix norm replacing the Frobenius norm above, giving the same minimizer $U_s\Sigma_s V_s^*$ but a different error than (B.4). For example, using the spectral norm $\|\cdot\|_2$, which returns the largest singular value of a matrix, gives the error

$$\|A - U_s\Sigma_s V_s^*\|_2 = \sigma_{s+1}, \tag{B.5}$$

which is the global minimum of the low-rank approximation problem.

B.4 ▪ Pseudoinverse

The SVD can also be used to generalize the notion of a matrix inverse to noninvertible and even nonsquare matrices. The notion of an inverse is generalized as follows: For a matrix $A \in \mathbb{C}^{M\times N}$, we seek a matrix $B \in \mathbb{C}^{N\times M}$ so that

1. $ABA = A$,
2. $BAB = B$,
3. $(AB)^* = AB$,
4. $(BA)^* = BA$.

Notice that the first condition above gives that AB is the left inverse of A and BA is its right inverse, which gives the criteria for B to be a weak inverse of A. Condition (2) similarly provides that A is a weak inverse of B, while (3) and (4) give that AB and BA are Hermitian matrices. The matrix B is called the pseudoinverse of A and is denoted $A^\dagger$. Moreover, if $A = U\Sigma V^*$ is the SVD of A, then $B = A^\dagger$ is given by

$$A^\dagger = V\Sigma^\dagger U^*, \tag{B.6}$$

where $\Sigma^\dagger$ is the pseudoinverse of Σ. Since $\Sigma \in \mathbb{C}^{M\times N}$ is a rectangular diagonal matrix, one can find that its pseudoinverse $\Sigma^\dagger \in \mathbb{C}^{N\times M}$ is again a rectangular diagonal matrix with diagonal entries given by

$$[\Sigma^\dagger]_{i,i} = \begin{cases} 1/\sigma_i, & \sigma_i > 0, \\ 0, & \sigma_i = 0, \end{cases} \tag{B.7}$$

where σ_i are the singular values of A. The result is that $\Sigma\Sigma^\dagger$ and $\Sigma^\dagger\Sigma$ closely resemble identity matrices since they are both square diagonal matrices with a 1 on the first $r = \operatorname{rank}(A)$ diagonal elements and zeros on the rest. With $A^\dagger$ in (B.6) we get the same result for $AA^\dagger$ and $A^\dagger A$.

Appendix C

Mathematical Programming

In Section 4 we were introduced to semidefinite programs (SDPs) for identifying auxiliary functions and invariant measures from data. Formulating and solving SDPs is a subfield of mathematical optimization that has found great use throughout much of applied mathematics and is now the subject of entire textbooks [93, 288]. Thus, in this appendix we only outline the basic components of linear, semidefinite, and sum-of-squares programs so that they can be recognized by the reader throughout this book. Computationally, these optimization methods can be formulated in MATLAB using the YALMIP package [177], for which tutorials can be found at https://yalmip.github.io/tutorials. YALMIP efficiently models these optimization problems but relies on external solvers to obtain their numerical solution, for which we use MOSEK [15].

C.1 ▪ Linear Programming

A **linear program** is a type of optimization problem for which the objective and all constraints are defined by linear relationships. The standard form of a linear program comes in the form of identifying a vector $x \in \mathbb{R}^d$ that solves the optimization problem

$$\max_{x\in\mathbb{R}^d} c\cdot x \quad \text{such that} \quad Ax \leq b \tag{C.1}$$

for fixed vectors $c \in \mathbb{R}^d$, $b \in \mathbb{R}^n$ and matrix $A \in \mathbb{R}^{n\times d}$. Here we have $n \geq 1$ linear inequality constraints enforced by the condition $Ax \leq b$, which is interpreted row-wise as

$$Ax \leq b \iff \begin{cases} a_{11}x_1 + \cdots + a_{1d}x_d \leq b_1, \\ \vdots \\ a_{n1}x_1 + \cdots + a_{nd}x_d \leq b_n, \end{cases} \tag{C.2}$$

with $x = (x_1, \dots, x_d)$ and a_{ij} the entries of matrix A. In the above and in what follows, some or all of the inequality conditions can be made equality conditions as well, but we refrain from this level of generalization here to keep the presentation concise.

The constraint $Ax \leq b$ defines the **feasible region** for the linear program (C.1). Geometrically, the feasible region is a convex polytope, generally with $(d-1)$-dimensional sides coming from forcing equality on one of the m constraints in (C.2). The objective function $c \cdot x$ can be thought of as a family of hyperplanes, and solving the linear program corresponds to translating the hyperplanes in the direction of c as much as one can while still intersecting the polytope. If the feasible region is bounded, then the point x that maximizes the objective function can be found on a vertex of the polytope.

C.2 ▪ Semidefinite Programming

A **semidefinite program** (SDP) generalizes the concept of linear programming to include semidefinite constraints on matrices. A general SDP can be written as

$$\max_{x \in \mathbb{R}^d} c \cdot x \quad \text{such that} \quad x_1 A_1 + \cdots + x_d A_d - B \succeq 0 \tag{C.3}$$

for a fixed vector $c \in \mathbb{R}^d$ and fixed *symmetric* matrices $A_1, \ldots, A_d, B \in \mathbb{R}^{m \times m}$. We may also consider (C.3) in the absence of the optimization condition $c \cdot x$, in which case one is simply testing for feasibility of the problem by searching for $x = (x_1, \ldots, x_d) \in \mathbb{R}^d$ so that $x_1 A_1 + \cdots + x_d A_d - B$ is a positive semidefinite matrix. Without an objective function, (C.3) is simply a **linear matrix inequality**, which was encountered in Section 3.3 for developing controllers to stabilize periodic orbits.

SDPs can be generalized to many forms beyond the standard presentation in (C.3). For example, one could instead consider the entries of the matrices of one or more of the matrices $A_1, \ldots, A_d$ to be variables and the $x_1, \ldots, x_d$ fixed. Thus, another general SDP formulation is

$$\max_{X \in \mathbb{S}^d} \operatorname{trace}(C^T X) \quad \text{such that} \quad X \succeq 0, \ \operatorname{trace}(A_k^T X) \leq b_k, \ k = 1, \ldots, m, \tag{C.4}$$

where $\mathbb{S}^d$ is the set of all symmetric matrices of size $d \times d$ and $C, A_1, \ldots, A_m \in \mathbb{S}^d$ with $m \geq 1$. Here the trace computations generalize the Euclidean dot product of vectors since

$$\operatorname{trace}(A^T B) = \sum_{i,j=1}^{d} A_{i,j} B_{i,j}, \tag{C.5}$$

simply representing the sum of the componentwise multiplications of the entries between A and B. In the case that matrices $C, A_1, \ldots, A_m$ are diagonal, (C.3) completely reduces to a linear program with unknowns given by the diagonal entries of X.

C.3 ▪ Sum-of-Squares Programming

The standard form of a sum-of-squares (SOS) optimization problem is

$$\max_{u \in \mathbb{R}^n} c \cdot u \quad \text{such that} \quad a_{1,j}(x) u_1 + \cdots + a_{n,j}(x) u_n \text{ is SOS } \forall j = 1, \ldots, J, \tag{C.6}$$

where $a_{i,j}(x)$ are polynomials for each $i = 1, \ldots, n$ and $j = 1, \ldots J$ for some $J \geq 1$. A function $f : \mathbb{R}^d \to \mathbb{R}$ is an SOS polynomial if and only if there exists a positive semidefinite matrix $P \succeq 0$ and a vector of monomials z in x so that $f(x) = z^\mathsf{T} P z$. Thus, the optimization problem (C.6) is equivalently formulated as an SDP in the form (C.3) by writing $a_{i,j}(x) = z^\mathsf{T} A_{i,j} z$ for symmetric $A_{i,j}$ to obtain the conditions $u_1 A_{1,j} + \cdots + u_d A_{d,j} \succeq 0$ for each $j = 1, \ldots, J$. This recasts (C.6) as an SDP of the form (C.3), while one may further instead take the coefficients of the polynomials $a_{i,j}(x)$ as variables to transform to (C.4). Identifying the matrices $A_{i,j}$ can be done automatically with YALMIP, leaving this translation to the SDP step to software.

Bibliography

[1] M. Abadi, A. Agarwal, P. Barham, E. Brevdo, Z. Chen, C. Citro, G. S. Corrado, A. Davis, J. Dean, M. Devin, et al. Tensorflow: Large-scale machine learning on heterogeneous distributed systems. *arXiv preprint arXiv:1603.04467*, 2016. (Cited on pp. 89, 96)

[2] B. Adcock, S. Brugiapaglia, N. Dexter, and S. Morage. Deep neural networks are effective at learning high-dimensional Hilbert-valued functions from limited data. In *Mathematical and Scientific Machine Learning*, pages 1–36. PMLR, 2022. (Cited on p. 107)

[3] B. Adcock, S. Brugiapaglia, and C. G. Webster. Compressed sensing approaches for polynomial approximation of high-dimensional functions. In *Compressed Sensing and Its Applications*, pages 93–124. Springer, 2017. (Cited on p. 107)

[4] B. Adcock, S. Brugiapaglia, and C. G. Webster. *Sparse Polynomial Approximation of High-Dimensional Functions*, Comput. Sci. Engrg. 25. SIAM, 2022. (Cited on p. 107)

[5] B. Adcock and N. Dexter. The gap between theory and practice in function approximation with deep neural networks. *SIAM Journal on Mathematics of Data Science*, 3(2):624–655, 2021. (Cited on p. 107)

[6] A. A. Ahmadi and B. El Khadir. Learning dynamical systems with side information. In *Learning for Dynamics and Control*, pages 718–727. PMLR, 2020. (Cited on p. 62)

[7] A. A. Ahmadi and A. Majumdar. DSOS and SDSOS optimization: More tractable alternatives to sum of squares and semidefinite optimization. *SIAM Journal on Applied Algebra and Geometry*, 3(2):193–230, 2019. (Cited on p. 68)

[8] A. A. Ahmadi, A. Olshevsky, P. A. Parrilo, and J. N. Tsitsiklis. NP-hardness of deciding convexity of quartic polynomials and related problems. *Mathematical Programming*, 137:453–476, 2013. (Cited on p. 68)

[9] A. A. Ahmadi and P. A. Parrilo. Non-monotonic Lyapunov functions for stability of discrete time nonlinear and switched systems. In *2008 47th IEEE Conference on Decision and Control*, pages 614–621. IEEE, 2008. (Cited on p. 76)

[10] S. Akshay, K. Soman, N. Mohan, and S. Sachin Kumar. Dynamic mode decomposition and its application in various domains: An overview. In *Applications in Ubiquitous Computing*, pages 121–132. Springer, 2021. (Cited on p. 35)

[11] A. Alassaf and L. Fan. Dynamic mode decomposition in various power system applications. In *2019 North American Power Symposium (NAPS)*, pages 1–6. IEEE, 2019. (Cited on p. 21)

[12] J. Alves and J. S. Ramos. Kneading theory for tree maps. *Ergodic Theory and Dynamical Systems*, 24(4):957–985, 2004. (Cited on p. 139)

[13] S.-i. Amari. Backpropagation and stochastic gradient descent method. *Neurocomputing*, 5(4–5):185–196, 1993. (Cited on p. 96)

[14] H. C. Andersen. Molecular dynamics simulations at constant pressure and/or temperature. *The Journal of Chemical Physics*, 72(4):2384–2393, 1980. (Cited on p. 16)

[15] M. ApS. *The MOSEK Optimization Toolbox for MATLAB Manual. Version 9.0*, 2019. (Cited on pp. 50, 69, 147)

[16] H. Arbabi and I. Mezić. Ergodic theory, dynamic mode decomposition, and computation of spectral properties of the Koopman operator. *SIAM Journal on Applied Dynamical Systems*, 16(4):2096–2126, 2017. (Cited on p. 21)

[17] L. Ardizzone, J. Kruse, C. Rother, and U. Köthe. Analyzing inverse problems with invertible neural networks. In *International Conference on Learning Representations*, 2018. (Cited on p. 114)

[18] V. I. Arnol'd. *Mathematical Methods of Classical Mechanics*, Grad. Texts in Math. 60. Springer Science & Business Media, 2013. (Cited on p. 128)

[19] T. Askham and J. N. Kutz. Variable projection methods for an optimized dynamic mode decomposition. *SIAM Journal on Applied Dynamical Systems*, 17(1):380–416, 2018. (Cited on p. 35)

[20] A. Avila and I. Mezić. Data-driven analysis and forecasting of highway traffic dynamics. *Nature Communications*, 11(1):1–16, 2020. (Cited on p. 21)

[21] P. J. Baddoo, B. Herrmann, B. J. McKeon, and S. L. Brunton. Kernel learning for robust dynamic mode decomposition: Linear and nonlinear disambiguation optimization. *Proceedings of the Royal Society A*, 478(2260):20210830, 2022. (Cited on p. 34)

[22] P. J. Baddoo, B. Herrmann, B. J. McKeon, J. Nathan Kutz, and S. L. Brunton. Physics-informed dynamic mode decomposition. *Proceedings of the Royal Society A*, 479(2271):20220576, 2023. (Cited on p. 16)

[23] J. Bakarji, K. Champion, J. Nathan Kutz, and S. L. Brunton. Discovering governing equations from partial measurements with deep delay autoencoders. *Proceedings of the Royal Society A*, 479(2276):20230422, 2023. (Cited on p. 139)

[24] A. G. Baydin, B. A. Pearlmutter, A. A. Radul, and J. M. Siskind. Automatic differentiation in machine learning: A survey. *Journal of Marchine Learning Research*, 18:1–43, 2018. (Cited on p. 99)

[25] G. Berkooz, P. Holmes, and J. L. Lumley. The proper orthogonal decomposition in the analysis of turbulent flows. *Annual Review of Fluid Mechanics*, 25(1):539–575, 1993. (Cited on pp. 5, 7)

[26] B. P. Bezruchko and D. A. Smirnov. *Extracting Knowledge from Time Series: An Introduction to Nonlinear Empirical Modeling*. Springer Science & Business Media, 2010. (Cited on p. 98)

[27] J. Billingham. Phase plane analysis of one-dimensional reaction diffusion waves with degenerate reaction terms. *Dynamics and Stability of Systems*, 15(1):23–33, 2000. (Cited on p. 71)

[28] J. Billingham and D. Needham. A note on the properties of a family of travelling-wave solutions arising in cubic autocatalysis. *Dynamics and Stability of Systems*, 6(1):33–49, 1991. (Cited on p. 71)

[29] J. Blechschmidt and O. G. Ernst. Three ways to solve partial differential equations with neural networks—a review. *GAMM-Mitteilungen*, 44(2):e202100006, 2021. (Cited on pp. 100, 107)

[30] G. Blekherman, P. A. Parrilo, and R. R. Thomas. *Semidefinite Optimization and Convex Algebraic Geometry*. SIAM, 2012. (Cited on p. 49)

[31] J. Bochi. Ergodic optimization of Birkhoff averages and Lyapunov exponents. In *Proceedings of the International Congress of Mathematicians 2018*, pages 1821–1842, Rio de Janeiro, 2018. (Cited on p. 87)

[32] L. Boninsegna, F. Nüske, and C. Clementi. Sparse learning of stochastic dynamical equations. *The Journal of Chemical Physics*, 148(24):241723, 2018. (Cited on p. 62)

[33] C. J. Bose and R. Murray. Minimum 'energy' approximations of invariant measures for nonsingular transformations. *Discrete and Continuous Dynamical Systems*, 14(3):597–615, 2006. (Cited on p. 82)

[34] C. J. Bose and R. Murray. Duality and the computation of approximate invariant densities for nonsingular transformations. *SIAM Journal on Optimization*, 18(2):691–709, 2007. (Cited on p. 82)

[35] G.-J. Both, G. Vermarien, and R. Kusters. Sparsely constrained neural networks for model discovery of PDEs. *arXiv preprint arXiv:2011.04336*, 2020. (Cited on p. 106)

[36] L. Bottou. Stochastic gradient descent tricks. In *Neural Networks: Tricks of the Trade*, pages 421–436. Springer, 2012. (Cited on p. 96)

[37] G. E. Box, G. M. Jenkins, G. C. Reinsel, and G. M. Ljung. *Time Series Analysis: Forecasting and Control.* John Wiley, 2015. (Cited on p. 21)

[38] J. J. Bramburger, S. L. Brunton, and J. N. Kutz. Deep learning of conjugate mappings. *Physica D: Nonlinear Phenomena*, 427:133008, 2021. (Cited on p. 135)

[39] J. J. Bramburger, S. Dahdah, and J. R. Forbes. Synthesizing control laws from data using sum-of-squares optimization. *arXiv preprint arXiv:2307.01089*, 2023. (Cited on p. 78)

[40] J. J. Bramburger, D. Dylewsky, and J. N. Kutz. Sparse identification of slow timescale dynamics. *Physical Review E*, 102(2):022204, 2020. (Cited on pp. 55, 58)

[41] J. J. Bramburger and G. Fantuzzi. Data-driven discovery of invariant measures. *Proceedings of the Royal Society A*, 480(2286):20230627, 2024. (Cited on pp. 82, 85, 87)

[42] J. J. Bramburger and G. Fantuzzi. Auxiliary functions as Koopman observables: Data-driven analysis of dynamical systems via polynomial optimization. *Journal of Nonlinear Science*, 34(1):8, 2024. (Cited on pp. 28, 75, 76, 78)

[43] J. J. Bramburger and D. Goluskin. Minimum wave speeds in monostable reaction–diffusion equations: Sharp bounds by polynomial optimization. *Proceedings of the Royal Society A*, 476(2241): 20200450, 2020. (Cited on p. 71)

[44] J. J. Bramburger and J. N. Kutz. Poincaré maps for multiscale physics discovery and nonlinear floquet theory. *Physica D: Nonlinear Phenomena*, 408:132479, 2020. (Cited on p. 44)

[45] J. J. Bramburger, J. N. Kutz, and S. L. Brunton. Data-driven stabilization of periodic orbits. *IEEE Access*, 9:43504–43521, 2021. (Cited on pp. 44, 51)

[46] R. Brasser, A. Barr, and V. Dobos. The tidal parameters of TRAPPIST-1b and c. *Monthly Notices of the Royal Astronomical Society*, 487(1):34–47, 2019. (Cited on p. 16)

[47] M. Brin and G. Stuck. *Introduction to Dynamical Systems.* Cambridge University Press, 2002. (Cited on p. 82)

[48] N. F. Britton et al. *Reaction-Diffusion Equations and Their Applications to Biology.* Academic Press, 1986. (Cited on p. 71)

[49] S. L. Brunton and J. N. Kutz. *Data-Driven Science and Engineering: Machine Learning, Dynamical Systems, and Control.* Cambridge University Press, 2019. (Cited on pp. 22, 44)

[50] S. L. Brunton, J. L. Proctor, and J. N. Kutz. Discovering governing equations from data by sparse identification of nonlinear dynamical systems. *Proceedings of the National Academy of Sciences*, 113(15):3932–3937, 2016. (Cited on p. 37)

[51] C. J. Burges. A tutorial on support vector machines for pattern recognition. *Data Mining and Knowledge Discovery*, 2(2):121–167, 1998. (Cited on p. 30)

[52] S. Cai, Z. Mao, Z. Wang, M. Yin, and G. E. Karniadakis. Physics-informed neural networks (PINNs) for fluid mechanics: A review. *Acta Mechanica Sinica*, 37(12):1727–1738, 2021. (Cited on pp. 99, 107)

[53] J. Carr. *Applications of Centre Manifold Theory*, Appl. Math. Sci. 35. Springer Science & Business Media, 2012. (Cited on p. 120)

[54] K. Champion, B. Lusch, J. N. Kutz, and S. L. Brunton. Data-driven discovery of coordinates and governing equations. *Proceedings of the National Academy of Sciences*, 116(45):22445–22451, 2019. (Cited on p. 135)

[55] Y. Chen, L. Lu, G. E. Karniadakis, and L. Dal Negro. Physics-informed neural networks for inverse problems in nano-optics and metamaterials. *Optics Express*, 28(8):11618–11633, 2020. (Cited on p. 105)

[56] Z. Chen, Y. Liu, and H. Sun. Physics-informed learning of governing equations from scarce data. *Nature Communications*, 12(1):1–13, 2021. (Cited on p. 106)

[57] A. Chernyavsky, J. J. Bramburger, G. Fantuzzi, and D. Goluskin. Convex relaxations of integral variational problems: Pointwise dual relaxation and sum-of-squares optimization. *SIAM Journal on Optimization*, 33(2):481–512, 2023. (Cited on p. 87)

[58] S. I. Chernyshenko, P. J. Goulart, D. Huang, and A. Papachristodoulou. Polynomial sum of squares in fluid dynamics: A review with a look ahead. *Philosophical Transactions of the Royal Society A*, 372(2020):20130350, 2014. (Cited on p. 87)

[59] M. J. Cho and R. H. Stockbridge. Linear programming formulation for optimal stopping problems. *SIAM Journal on Control and Optimization*, 40(6):1965–1982, 2002. (Cited on p. 87)

[60] E. A. Christensen, M. Brøns, and J. N. Sørensen. Evaluation of proper orthogonal decomposition–based decomposition techniques applied to parameter-dependent nonturbulent flows. *SIAM Journal on Scientific Computing*, 21(4):1419–1434, 2000. (Cited on p. 7)

[61] D. S. Cohen, J. C. Neu, and R. R. Rosales. Rotating spiral wave solutions of reaction-diffusion equations. *SIAM Journal on Applied Mathematics*, 35(3):536–547, 1978. (Cited on p. 7)

[62] R. Collobert and J. Weston. A unified architecture for natural language processing: Deep neural networks with multitask learning. In *Proceedings of the 25th International Conference on Machine Learning*, pages 160–167, 2008. (Cited on p. 107)

[63] P. Conti, G. Gobat, S. Fresca, A. Manzoni, and A. Frangi. Reduced order modeling of parametrized systems through autoencoders and SINDy approach: Continuation of periodic solutions. *Computer Methods in Applied Mechanics and Engineering*, 411:116072, 2023. (Cited on p. 139)

[64] F. Covella and G. Fantuzzi. Uncertainty propagation for nonlinear dynamics: A polynomial optimization approach. In *2023 American Control Conference (ACC)*, pages 4142–4147. IEEE, 2023. (Cited on p. 87)

[65] N. Črnjarić-Žic, S. Maćešić, and I. Mezić. Koopman operator spectrum for random dynamical systems. *Journal of Nonlinear Science*, 30:2007–2056, 2020. (Cited on p. 74)

[66] S. Cuomo, V. S. Di Cola, F. Giampaolo, G. Rozza, M. Raissi, and F. Piccialli. Scientific machine learning through physics-informed neural networks: Where we are and what's next. *Journal of Scientific Computing*, 92(3):88, 2022. (Cited on pp. 99, 107)

[67] P. Cvitanovic, R. Artuso, R. Mainieri, G. Tanner, G. Vattay, N. Whelan, and A. Wirzba. Chaos: Classical and quantum. *ChaosBook. org (Niels Bohr Institute, Copenhagen 2005)*, 69:25, 2005. (Cited on p. 138)

[68] G. Cybenko. Approximation by superpositions of a sigmoidal function. *Mathematics of Control, Signals and Systems*, 2(4):303–314, 1989. (Cited on pp. 90, 91)

[69] G. Davis, S. Mallat, and M. Avellaneda. Adaptive greedy approximations. *Constructive Approximation*, 13(1):57–98, 1997. (Cited on p. 38)

[70] B. de Silva, K. Champion, M. Quade, J.-C. Loiseau, J. N. Kutz, and S. Brunton. PySINDy: A Python package for the sparse identification of nonlinear dynamical systems from data. *Journal of Open Source Software*, 5(49):1–4, 2020. (Cited on p. 62)

[71] D. Della-Bosca, D. Patterson, and S. Costain. Fractal complexity in built and game environments. In *Entertainment Computing–ICEC 2014: 13th International Conference, ICEC 2014, Sydney, Australia, October 1–3, 2014. Proceedings 13*, pages 167–172. Springer, 2014. (Cited on p. 4)

[72] M. Dellnitz and O. Junge. On the approximation of complicated dynamical behavior. *SIAM Journal on Numerical Analysis*, 36(2):491–515, 1999. (Cited on p. 87)

[73] R. L. Devaney. *An Introduction to Chaotic Dynamical Systems*. CRC Press, 2018. (Cited on pp. 111, 139)

[74] T. A. Driscoll, N. Hale, and L. N. Trefethen, eds. *Chebfun Guide*, Pafnuty Publications, 2014. (Cited on p. 77)

[75] G. E. Dullerud and F. Paganini. *A Course in Robust Control Theory: A Convex Approach*, Texts in Appl. Math. 36. Springer Science & Business Media, 2013. (Cited on p. 48)

[76] D. Dylewsky, E. Kaiser, S. L. Brunton, and J. N. Kutz. Principal component trajectories for modeling spectrally continuous dynamics as forced linear systems. *Physical Review E*, 105(1):015312, 2022. (Cited on p. 21)

[77] D. Dylewsky, M. Tao, and J. N. Kutz. Dynamic mode decomposition for multiscale nonlinear physics. *Physical Review E*, 99(6):063311, 2019. (Cited on pp. 16, 17, 18, 19, 57)

[78] C. Eckart and G. Young. The approximation of one matrix by another of lower rank. *Psychometrika*, 1(3):211–218, 1936. (Cited on p. 144)

[79] D. Elbrächter, D. Perekrestenko, P. Grohs, and H. Bölcskei. Deep neural network approximation theory. *IEEE Transactions on Information Theory*, 67(5):2581–2623, 2021. (Cited on p. 107)

[80] B. Engquist. *Encyclopedia of Applied and Computational Mathematics*. Springer, 2015. (Cited on p. 140)

[81] S. A. Falconer, D. J. Lloyd, and N. Santitissadeekorn. Combining dynamic mode decomposition with ensemble Kalman filtering for tracking and forecasting. *Physica D: Nonlinear Phenomena*, 449:133741, 2023. (Cited on p. 21)

[82] G. Fantuzzi. Verification of some functional inequalities via polynomial optimization. *IFAC-PapersOnLine*, 55(16):166–171, 2022. (Cited on p. 87)

[83] G. Fantuzzi, D. Goluskin, D. Huang, and S. I. Chernyshenko. Bounds for deterministic and stochastic dynamical systems using sum-of-squares optimization. *SIAM Journal on Applied Dynamical Systems*, 15(4):1962–1988, 2016. (Cited on p. 87)

[84] G. Fantuzzi and I. Tobasco. Sharpness and non-sharpness of occupation measure bounds for integral variational problems. *arXiv preprint arXiv:2207.13570*, 2022. (Cited on p. 87)

[85] C. L. Fefferman. Existence and smoothness of the Navier–Stokes equation. *The Millennium Prize Problems*, 57:67, 2000. (Cited on p. 2)

[86] M. J. Feigenbaum. Quantitative universality for a class of nonlinear transformations. *Journal of Statistical Physics*, 19(1):25–52, 1978. (Cited on p. 112)

[87] P. C. Fife and J. B. McLeod. The approach of solutions of nonlinear diffusion equations to travelling front solutions. *Archive for Rational Mechanics and Analysis*, 65(4):335–361, 1977. (Cited on pp. 102, 103)

[88] G. Floquet. Sur les équations différentielles linéaires à coefficients périodiques. *Annales Scientifiques de l'École Normale Supérieure*, 12:47–88, 1883. (Cited on p. 44)

[89] D. Floryan and M. D. Graham. Data-driven discovery of intrinsic dynamics. *Nature Machine Intelligence*, 4(12):1113–1120, 2022. (Cited on p. 139)

[90] C. Folkestad, D. Pastor, I. Mezic, R. Mohr, M. Fonoberova, and J. Burdick. Extended dynamic mode decomposition with learned Koopman eigenfunctions for prediction and control. In *2020 American Control Conference (ACC)*, pages 3906–3913. IEEE, 2020. (Cited on pp. 25, 26)

[91] E. Forest. Geometric integration for particle accelerators. *Journal of Physics A: Mathematical and General*, 39(19):5321, 2006. (Cited on p. 57)

[92] P. Frederickson, J. L. Kaplan, E. D. Yorke, and J. A. Yorke. The Liapunov dimension of strange attractors. *Journal of Differential Equations*, 49(2):185–207, 1983. (Cited on p. 136)

[93] B. Gärtner and J. Matousek. *Approximation algorithms and semidefinite programming*. Springer Science & Business Media, 2012. (Cited on p. 147)

[94] S. Gerber, S. Olsson, F. Noé, and I. Horenko. A scalable approach to the computation of invariant measures for high-dimensional Markovian systems. *Scientific Reports*, 8(1):1796, 2018. (Cited on p. 87)

[95] C. Gin, B. Lusch, S. L. Brunton, and J. N. Kutz. Deep learning models for global coordinate transformations that linearise PDEs. *European Journal of Applied Mathematics*, 32(3):515–539, 2021. (Cited on pp. 28, 123, 127)

[96] C. Gissinger. A new deterministic model for chaotic reversals. *The European Physical Journal B*, 85(4):1–12, 2012. (Cited on p. 136)

[97] P. A. Glendinning and D. J. W. Simpson. Normal forms, differentiable conjugacies, and elementary bifurcations of maps. *SIAM Journal on Applied Mathematics*, 83(2):816–836, 2023. (Cited on pp. 117, 120, 123)

[98] E. Goceri. Analysis of deep networks with residual blocks and different activation functions: Classification of skin diseases. In *2019 Ninth International Conference on Image Processing Theory, Tools and Applications (IPTA)*, pages 1–6. IEEE, 2019. (Cited on p. 103)

[99] M. P. Golden and B. E. Ydstie. Bifurcation in model reference adaptive control systems. *Systems & Control Letters*, 11(5):413–430, 1988. (Cited on p. 121)

[100] G. H. Golub and C. F. Van Loan. *Matrix Computations*. Johns Hopkins University Press, 2013. (Cited on p. 12)

[101] D. Goluskin. Bounding averages rigorously using semidefinite programming: Mean moments of the Lorenz system. *Journal of Nonlinear Science*, 28(2):621–651, 2018. (Cited on p. 67)

[102] D. Goluskin. Bounding extrema over global attractors using polynomial optimisation. *Nonlinearity*, 33(9):4878, 2020. (Cited on p. 67)

[103] D. Goluskin and G. Fantuzzi. Bounds on mean energy in the Kuramoto–Sivashinsky equation computed using semidefinite programming. *Nonlinearity*, 32(5):1705, 2019. (Cited on p. 87)

[104] I. Goodfellow, Y. Bengio, and A. Courville. *Deep Learning*. MIT Press, 2016. (Cited on p. 95)

[105] D. Goswami, E. Thackray, and D. A. Paley. Constrained Ulam dynamic mode decomposition: Approximation of the Perron-Frobenius operator for deterministic and stochastic systems. *IEEE Control Systems Letters*, 2(4):809–814, 2018. (Cited on p. 87)

[106] T. Gowers, J. Barrow-Green, and I. Leader. *The Princeton Companion to Mathematics*. Princeton University Press, 2008. (Cited on p. 43)

[107] J. Grosek and J. N. Kutz. Dynamic mode decomposition for real-time background/foreground separation in video. *arXiv preprint arXiv:1404.7592*, 2014. (Cited on pp. 13, 35)

[108] J. Guckenheimer. Sensitive dependence to initial conditions for one dimensional maps. *Communications in Mathematical Physics*, 70(2):133–160, 1979. (Cited on p. 132)

[109] J. Guckenheimer and P. Holmes. *Nonlinear Oscillations, Dynamical Systems, and Bifurcations of Vector Fields*, Appl. Math. Sci. 42. Springer Science & Business Media, 2013. (Cited on pp. 3, 22, 45, 54, 112, 117, 120)

[110] M. Guo, C. De Persis, and P. Tesi. Learning control for polynomial systems using sum of squares relaxations. In *2020 59th IEEE Conference on Decision and Control (CDC)*, pages 2436–2441. IEEE, 2020. (Cited on p. 78)

[111] J. H. Tu, C. W. Rowley, D. M. Luchtenburg, S. L. Brunton, and J. Nathan Kutz. On dynamic mode decomposition: Theory and applications. *Journal of Computational Dynamics*, 1(2):391–421, 2014. (Cited on pp. 16, 22, 34)

[112] E. Haghighat, M. Raissi, A. Moure, H. Gomez, and R. Juanes. A deep learning framework for solution and discovery in solid mechanics: Linear elasticity. *arXiv preprint arXiv:2003.02751*, 2020. (Cited on p. 105)

[113] J. Hale. *Ordinary Differential Equations*. Dover Books on Mathematics Series. Dover Publications, 2009. (Cited on p. 55)

[114] I. U. Haq, K. Fujii, and Y. Kawahara. Dynamic mode decomposition via dictionary learning for foreground modeling in videos. *Computer Vision and Image Understanding*, 199:103022, 2020. (Cited on p. 35)

[115] M. H. Hassoun et al. *Fundamentals of Artificial Neural Networks*. MIT Press, 1995. (Cited on p. 91)

[116] P. Heas and C. Herzet. State-of-the-art algorithms for low-rank dynamic mode decomposition. *arXiv preprint arXiv:2108.09160*, 2021. (Cited on p. 34)

[117] L. H. Hellström and A. J. Smits. Structure identification in pipe flow using proper orthogonal decomposition. *Philosophical Transactions of the Royal Society A: Mathematical, Physical and Engineering Sciences*, 375(2089):20160086, 2017. (Cited on p. 7)

[118] M. Hénon. A two-dimensional mapping with a strange attractor. *Communications in Mathematical Physics*, 50(1):69–77, 1976. (Cited on p. 97)

[119] D. Henrion. Semidefinite characterisation of invariant measures for one-dimensional discrete dynamical systems. *Kybernetika*, 48(6):1089–1099, 2012. (Cited on p. 83)

[120] D. Henrion and A. Garulli. *Positive Polynomials in Control*, Lect. Notes Control Inform. Sci. 312. Springer Science & Business Media, 2005. (Cited on p. 78)

[121] D. Henrion, M. Korda, M. Kružík, and R. Rios-Zertuche. Occupation measure relaxations in variational problems: The role of convexity. *SIAM Journal on Optimization*, 34(2):1708–1731, 2024. (Cited on p. 87)

[122] D. Henrion, J. B. Lasserre, and C. Savorgnan. Nonlinear optimal control synthesis via occupation measures. In *Proceedings of the IEEE Conference on Decision & Control*, pages 4749–4754, 2008. IEEE. (Cited on p. 87)

[123] D. Hernández-Hernández, O. Hernández-Lerma, and M. Taksar. The linear programming approach to deterministic optimal control problems. *Applicationes Mathematicae*, 24(1):17–33, 1996. (Cited on p. 87)

[124] D. Hilbert. Über die darstellung definiter formen als summe von formenquadraten. In *Algebra· Invariantentheorie Geometrie*, pages 154–161. Springer, 1970. (Cited on p. 69)

[125] N. Hitchin, G. Segal, and R. Ward. *Integrable Systems: Twistors, Loop Groups, and Riemann Surfaces*. Oxford Graduate Texts in Mathematics. Oxford University Press, 2013. (Cited on p. 58)

[126] M. Hoffmann, C. Fröhner, and F. Noé. Reactive SINDy: Discovering governing reactions from concentration data. *The Journal of Chemical Physics*, 150(2):025101, 2019. (Cited on p. 62)

[127] K. Hornik. Approximation capabilities of multilayer feedforward networks. *Neural Networks*, 4(2):251–257, 1991. (Cited on p. 91)

[128] Z. Hu, A. D. Jagtap, G. E. Karniadakis, and K. Kawaguchi. When do extended physics-informed neural networks (XPINNs) improve generalization? *SIAM Journal on Scientific Computing*, 44(5):A3158–A3182, 2022. (Cited on p. 107)

[129] I. Ishikawa, T. Teshima, K. Tojo, K. Oono, M. Ikeda, and M. Sugiyama. Universal approximation property of invertible neural networks. *Journal of Machine Learning Research*, 24(287):1–68, 2023. (Cited on p. 114)

[130] A. D. Jagtap and G. E. Karniadakis. Extended physics-informed neural networks (XPINNs): A generalized space-time domain decomposition based deep learning framework for nonlinear partial differential equations. In *AAAI Spring Symposium: MLPS*, 2021. (Cited on p. 107)

[131] W. Ji, W. Qiu, Z. Shi, S. Pan, and S. Deng. Stiff-PINN: Physics-informed neural network for stiff chemical kinetics. *The Journal of Physical Chemistry A*, 125(36):8098–8106, 2021. (Cited on p. 107)

[132] Y. Jiao, Y. Lai, D. Li, X. Lu, F. Wang, Y. Wang, and J. Z. Yang. A rate of convergence of physics informed neural networks for the linear second order elliptic PDEs. *Communications in Computational Physics*, 31(4):1272, 2022. (Cited on p. 107)

[133] I. T. Jolliffe and J. Cadima. Principal component analysis: A review and recent developments. *Philosophical Transactions of the Royal Society A: Mathematical, Physical and Engineering Sciences*, 374(2065):20150202, 2016. (Cited on p. 5)

[134] M. Jones and M. M. Peet. Using SOS and sublevel set volume minimization for estimation of forward reachable sets. *IFAC-PapersOnLine*, 52(16):484–489, 2019. (Cited on p. 87)

[135] K. Kaheman, J. N. Kutz, and S. L. Brunton. SINDy-PI: A robust algorithm for parallel implicit sparse identification of nonlinear dynamics. *Proceedings of the Royal Society A*, 476(2242):20200279, 2020. (Cited on p. 62)

[136] E. Kaiser, J. N. Kutz, and S. L. Brunton. Discovering conservation laws from data for control. In *2018 IEEE Conference on Decision and Control (CDC)*, pages 6415–6421. IEEE, 2018. (Cited on pp. 59, 60)

[137] M. Kalia, S. L. Brunton, H. G. Meijer, C. Brune, and J. N. Kutz. Deep learning of normal form autoencoders for universal, parameter-dependent dynamics. In *First NeurIPS Workshop on Interpretable Inductive Biases and Physically Structured Learning*, 2020. (Cited on p. 118)

[138] J. L. Kaplan and J. A. Yorke. Chaotic behavior of multidimensional difference equations. In *Functional Differential Equations and Approximation of Fixed Points*, pages 204–227. Springer, 1979. (Cited on p. 136)

[139] J. Kappraff. Complexity and chaos theory in art. In *On Art and Science: Tango of an Eternally Inseparable Duo*, pages 117–138. Springer, 2019. (Cited on p. 4)

[140] A. A. Kaptanoglu, J. L. Callaham, A. Aravkin, C. J. Hansen, and S. L. Brunton. Promoting global stability in data-driven models of quadratic nonlinear dynamics. *Physical Review Fluids*, 6(9):094401, 2021. (Cited on p. 62)

[141] A. A. Kaptanoglu, B. M. de Silva, U. Fasel, K. Kaheman, A. J. Goldschmidt, J. Callaham, C. B. Delahunt, Z. G. Nicolaou, K. Champion, J.-C. Loiseau, et al. PySINDy: A comprehensive Python package for robust sparse system identification. *Journal of Open Source Software*, 7(69):3994, 2022. (Cited on p. 62)

[142] M. Karplus and J. A. McCammon. Molecular dynamics simulations of biomolecules. *Nature Structural Biology*, 9(9):646–652, 2002. (Cited on p. 16)

[143] T. H. Keitt, M. A. Lewis, and R. D. Holt. Allee effects, invasion pinning, and species' borders. *The American Naturalist*, 157(2):203–216, 2001. (Cited on p. 102)

[144] P. Kidger and T. Lyons. Universal approximation with deep narrow networks. In *Conference on Learning Theory*, pages 2306–2327. PMLR, 2020. (Cited on p. 92)

[145] D. P. Kingma and J. Ba. Adam: A method for stochastic optimization. *arXiv preprint arXiv:1412.6980*, 2014. (Cited on p. 96)

[146] G. Klambauer, T. Unterthiner, A. Mayr, and S. Hochreiter. Self-normalizing neural networks. In *Advances in Neural Information Processing Systems*, 30, 2017. (Cited on p. 103)

[147] S. Klus, F. Nüske, S. Peitz, J.-H. Niemann, C. Clementi, and C. Schütte. Data-driven approximation of the Koopman generator: Model reduction, system identification, and control. *Physica D: Nonlinear Phenomena*, 406:132416, 2020. (Cited on pp. 74, 87)

[148] S. Klus and C. Schütte. Towards tensor-based methods for the numerical approximation of the Perron-Frobenius and Koopman operator. *Journal of Computational Dynamics*, 3(2):139–161, 2016. (Cited on p. 87)

[149] E. Knobloch. On the statistical dynamics of the Lorenz model. *Journal of Statistical Physics*, 20(6):695–709, 1979. (Cited on p. 67)

[150] B. O. Koopman. Hamiltonian systems and transformation in Hilbert space. *Proceedings of the National Academy of Sciences of the United States of America*, 17(5):315, 1931. (Cited on p. 25)

[151] M. Korda, D. Henrion, and C. N. Jones. Convex computation of the maximum controlled invariant set for polynomial control systems. *SIAM Journal on Control and Optimization*, 52(5):2944–2969, 2014. (Cited on p. 87)

[152] M. Korda, D. Henrion, and J. B. Lasserre. Moments and convex optimization for analysis and control of nonlinear pdes. In *Numerical Control: Part A*, Handb. Numer. Anal. 23, pages 339–366. Elsevier, 2022. (Cited on p. 87)

[153] M. Korda, D. Henrion, and I. Mezić. Convex computation of extremal invariant measures of nonlinear dynamical systems and Markov processes. *Journal of Nonlinear Science*, 31:1–26, 2021. (Cited on pp. 82, 83, 85)

[154] M. Korda and I. Mezić. On convergence of extended dynamic mode decomposition to the Koopman operator. *Journal of Nonlinear Science*, 28(2):687–710, 2018. (Cited on p. 28)

[155] M. A. Kramer. Nonlinear principal component analysis using autoassociative neural networks. *AIChE Journal*, 37(2):233–243, 1991. (Cited on p. 114)

[156] N. Kryloff and N. Bogoliouboff. La théorie générale de la mesure dans son application à l'étude des systèmes dynamiques de la mécanique non linéaire. *Annals of Mathematics* (2), 38(1):65–113, 1937. (Cited on p. 82)

[157] S. Kundu and M. Anghel. A sum-of-squares approach to the stability and control of interconnected systems using vector Lyapunov functions. In *2015 American Control Conference (ACC)*, pages 5022–5028. IEEE, 2015. (Cited on p. 78)

[158] J. Kuntz, M. Ottobre, G.-B. Stan, and M. Barahona. Bounding stationary averages of polynomial diffusions via semidefinite programming. *SIAM Journal on Scientific Computing*, 38(6):A3891–A3920, 2016. (Cited on p. 66)

[159] J. Kuntz, P. Thomas, G.-B. Stan, and M. Barahona. Bounding the stationary distributions of the chemical master equation via mathematical programming. *The Journal of Chemical Physics*, 151(3):034109, 2019. (Cited on p. 66)

[160] J. N. Kutz. *Data-Driven Modeling and Scientific Computation: Methods for Complex Systems and Big Data*. Oxford University Press, 2013. (Cited on pp. 7, 13, 17)

[161] J. N. Kutz, A. Rahman, M. R. Ebers, J. Koch, and J. J. Bramburger. Universal dynamics of damped-driven systems: The logistic map as a normal form for energy balance. *arXiv preprint arXiv:2211.11748*, 2022. (Cited on p. 135)

[162] Y. A. Kuznetsov, I. A. Kuznetsov, and Y. Kuznetsov. *Elements of Applied Bifurcation Theory*, Appl. Math. Sci. 112. Springer, 1998. (Cited on pp. 3, 117, 120, 121)

[163] J. P. La Salle. *The Stability of Dynamical Systems*. SIAM, 1976. (Cited on p. 64)

[164] I. E. Lagaris, A. Likas, and D. I. Fotiadis. Artificial neural networks for solving ordinary and partial differential equations. *IEEE Transactions on Neural Networks*, 9(5):987–1000, 1998. (Cited on p. 99)

[165] Y. Lan and I. Mezić. Linearization in the large of nonlinear systems and Koopman operator spectrum. *Physica D: Nonlinear Phenomena*, 242(1):42–53, 2013. (Cited on p. 113)

[166] H. Lange, S. L. Brunton, and J. N. Kutz. From Fourier to Koopman: Spectral methods for long-term time series prediction. *Journal of Machine Learning Research*, 22(41):1–38, 2021. (Cited on p. 139)

[167] J. Laskar. Secular evolution of the solar system over 10 million years. *Astronomy and Astrophysics*, 198:341–362, 1988. (Cited on p. 16)

[168] A. Lasota and M. C. Mackey. *Chaos, Fractals, and Noise: Stochastic Aspects of Dynamics*, Appl. Math. Sci. 97. Springer Science & Business Media, 1998. (Cited on p. 85)

[169] J. B. Lasserre. A sum of squares approximation of nonnegative polynomials. *SIAM Review*, 49(4):651–669, 2007. (Cited on p. 69)

[170] J. B. Lasserre, D. Henrion, C. Prieur, and E. Trélat. Nonlinear optimal control via occupation measures and LMI-relaxations. *SIAM Journal on Control and Optimization*, 47(4):1643–1666, 2008. (Cited on p. 87)

[171] M. Laurent. Sums of squares, moment matrices and optimization over polynomials. In *Emerging Applications of Algebraic Geometry*, pages 157–270. Springer, 2009. (Cited on p. 86)

[172] M. Lee and J. Park. An optimized dynamic mode decomposition model robust to multiplicative noise. *SIAM Journal on Applied Dynamical Systems*, 22(1):235–268, 2023. (Cited on p. 35)

[173] Q. Li, F. Dietrich, E. M. Bollt, and I. G. Kevrekidis. Extended dynamic mode decomposition with dictionary learning: A data-driven adaptive spectral decomposition of the Koopman operator. *Chaos: An Interdisciplinary Journal of Nonlinear Science*, 27(10):103111, 2017. (Cited on p. 28)

[174] T.-Y. Li and J. A. Yorke. Period three implies chaos. *The American Mathematical Monthly*, 82(10):985–992, 1975. (Cited on p. 117)

[175] A. J. Linot and M. D. Graham. Deep learning to discover and predict dynamics on an inertial manifold. *Physical Review E*, 101(6):062209, 2020. (Cited on p. 139)

[176] A. J. Linot and M. D. Graham. Data-driven reduced-order modeling of spatiotemporal chaos with neural ordinary differential equations. *Chaos: An Interdisciplinary Journal of Nonlinear Science*, 32(7):073110, 2022. (Cited on p. 139)

[177] J. Lofberg. YALMIP: A toolbox for modeling and optimization in MATLAB. In *2004 IEEE International Conference on Robotics and Automation (IEEE Cat. No. 04CH37508)*, pages 284–289. IEEE, 2004. (Cited on pp. 50, 70, 147)

[178] E. N. Lorenz. Deterministic nonperiodic flow. *Journal of Atmospheric Sciences*, 20(2):130–141, 1963. (Cited on p. 1)

[179] Z. Lu, H. Pu, F. Wang, Z. Hu, and L. Wang. The expressive power of neural networks: A view from the width. In *Advances in Neural Information Processing Systems*, 30, 2017. (Cited on p. 92)

[180] B. Lusch, J. N. Kutz, and S. L. Brunton. Deep learning for universal linear embeddings of nonlinear dynamics. *Nature Communications*, 9(1):1–10, 2018. (Cited on pp. 28, 123)

[181] A. M. Lyapunov. The general problem of the stability of motion. *International Journal of Control*, 55(3):531–534, 1992. (Cited on p. 63)

[182] B. Macukow. Neural networks—State of art, brief history, basic models and architecture. In *IFIP International Conference on Computer Information Systems and Industrial Management*, pages 3–14. Springer, 2016. (Cited on p. 89)

[183] T. Maeba, M. Deng, A. Yanou, and T. Henmi. Swing-up controller design for inverted pendulum by using energy control method based on Lyapunov function. In *Proceedings of the 2010 International Conference on Modelling, Identification and Control*, pages 768–773. IEEE, 2010. (Cited on p. 80)

[184] V. Magron, M. Forets, and D. Henrion. Semidefinite approximations of invariant measures for polynomial systems. *Discrete and Continuous Dynamical Systems Series B*, 22(11):1–26, 2017. (Cited on p. 87)

[185] V. Magron, P.-L. Garoche, D. Henrion, and X. Thirioux. Semidefinite approximations of reachable sets for discrete-time polynomial systems. *SIAM Journal on Control and Optimization*, 57(4):2799–2820, 2019. (Cited on p. 87)

[186] R. Malhotra, M. Holman, and T. Ito. Chaos and stability of the solar system. *Proceedings of the National Academy of Sciences*, 98(22):12342–12343, 2001. (Cited on p. 16)

[187] R. Mané. *Ergodic Theory and Differentiable Dynamics*, MATHE3 8. Springer Science & Business Media, 2012. (Cited on p. 66)

[188] N. M. Mangan, J. N. Kutz, S. L. Brunton, and J. L. Proctor. Model selection for dynamical systems via sparse regression and information criteria. *Proceedings of the Royal Society A: Mathematical, Physical and Engineering Sciences*, 473(2204):20170009, 2017. (Cited on p. 44)

[189] J. Mann and J. N. Kutz. Dynamic mode decomposition for financial trading strategies. *Quantitative Finance*, 16(11):1643–1655, 2016. (Cited on p. 35)

[190] D. G. Martinson. *Empirical Orthogonal Function (EOF) Analysis*, pages 495–534. Cambridge University Press, 2018. (Cited on p. 5)

[191] A. Mauroy and J. Goncalves. Koopman-based lifting techniques for nonlinear systems identification. *IEEE Transactions on Automatic Control*, 65(6):2550–2565, 2019. (Cited on p. 36)

[192] W. S. McCulloch and W. Pitts. A logical calculus of the ideas immanent in nervous activity. *The Bulletin of Mathematical Biophysics*, 5(4):115–133, 1943. (Cited on p. 89)

[193] H. McKean, Jr. Nagumo's equation. *Advances in Mathematics*, 4(3):209–223, 1970. (Cited on p. 101)

[194] D. A. Messenger and D. M. Bortz. Weak SINDy for partial differential equations. *Journal of Computational Physics*, 443:110525, 2021. (Cited on p. 41)

[195] D. A. Messenger and D. M. Bortz. Weak SINDy: Galerkin-based data-driven model selection. *Multiscale Modeling & Simulation*, 19(3):1474–1497, 2021. (Cited on p. 41)

[196] J. Miller, D. Henrion, M. Sznaier, and M. Korda. Peak estimation for uncertain and switched systems. In *Proceedings of the 60th IEEE Conference on Decision and Control*, pages 3222–3228, 2021. (Cited on p. 87)

[197] J. Milnor and W. Thurston. On iterated maps of the interval. In *Dynamical Systems*, pages 465–563. Springer, 1988. (Cited on pp. 112, 132, 139)

[198] L. Mirsky. Symmetric gauge functions and unitarily invariant norms. *The Quarterly Journal of Mathematics*, 11(1):50–59, 1960. (Cited on p. 144)

[199] S. Mishra and R. Molinaro. Estimates on the generalization error of physics-informed neural networks for approximating a class of inverse problems for PDEs. *IMA Journal of Numerical Analysis*, 42(2):981–1022, 2022. (Cited on p. 107)

[200] G. S. Misyris, A. Venzke, and S. Chatzivasileiadis. Physics-informed neural networks for power systems. In *2020 IEEE Power & Energy Society General Meeting (PESGM)*, pages 1–5. IEEE, 2020. (Cited on p. 105)

[201] F. K. Moore and E. M. Greitzer. A theory of post-stall transients in axial compression systems: Part I—Development of equations. *Journal of Engineering for Gas Turbines and Power*, 108(1):68–76, 01 1986. (Cited on p. 69)

[202] T. S. Motzkin. The arithmetic-geometric inequality. In *Inequalities (Proceedings of a Symposium Held at Wright-Patterson Air Force Base, Ohio, 1965)*, pages 205–224, 1967. (Cited on p. 69)

[203] J. A. Murdock. *Perturbations: Theory and Methods*. SIAM, 1999. (Cited on p. 54)

[204] K. G. Murty and S. N. Kabadi. Some NP-complete problems in quadratic and nonlinear programming. *Mathematical Programming*, 39:117–129, 1987. (Cited on p. 68)

[205] B. K. Natarajan. Sparse approximate solutions to linear systems. *SIAM Journal on Computing*, 24(2):227–234, 1995. (Cited on p. 38)

[206] S. E. Newhouse, J. Palis, and F. Takens. Bifurcations and stability of families of diffeomorphisms. *Publications Mathématiques de l'IHÉS*, 57:5–71, 1983. (Cited on pp. 117, 118, 120)

[207] S. Nikolopoulos, I. Kalogeris, and V. Papadopoulos. Non-intrusive surrogate modeling for parametrized time-dependent partial differential equations using convolutional autoencoders. *Engineering Applications of Artificial Intelligence*, 109:104652, 2022. (Cited on p. 107)

[208] T. Nonomura, H. Shibata, and R. Takaki. Dynamic mode decomposition using a Kalman filter for parameter estimation. *AIP Advances*, 8(10):105106, 2018. (Cited on p. 35)

[209] V. I. Oseledec. A multiplicative ergodic theorem: Liapunov characteristic number for dynamical systems. *Transactions of the Moscow Mathematical Society*, 19:197–231, 1968. (Cited on p. 135)

[210] E. Ott, C. Grebogi, and J. A. Yorke. Controlling chaos. *Physical Review Letters*, 64(11):1196, 1990. (Cited on p. 48)

[211] J. Page and R. R. Kerswell. Koopman analysis of Burgers equation. *Physical Review Fluids*, 3(7):071901, 2018. (Cited on pp. 32, 33, 34)

[212] M. Paluš. Multiscale atmospheric dynamics: Cross-frequency phase-amplitude coupling in the air temperature. *Physical Review Letters*, 112(7):078702, 2014. (Cited on p. 16)

[213] S. Pan and K. Duraisamy. On the structure of time-delay embedding in linear models of non-linear dynamical systems. *Chaos: An Interdisciplinary Journal of Nonlinear Science*, 30(7):073135, 2020. (Cited on pp. 21, 22)

[214] G. Pang, M. D'Elia, M. Parks, and G. E. Karniadakis. nPINNs: Nonlocal physics-informed neural networks for a parametrized nonlocal universal Laplacian operator. Algorithms and applications. *Journal of Computational Physics*, 422:109760, 2020. (Cited on p. 107)

[215] Y. Pantazis and I. Tsamardinos. A unified approach for sparse dynamical system inference from temporal measurements. *Bioinformatics*, 35(18):3387–3396, 2019. (Cited on p. 41)

[216] A. Papachristodoulou and S. Prajna. A tutorial on sum of squares techniques for systems analysis. In *Proceedings of the 2005 American Control Conference, 2005*, pages 2686–2700. IEEE, 2005. (Cited on p. 78)

[217] J. P. Parker, D. Goluskin, and G. M. Vasil. A study of the double pendulum using polynomial optimization. *Chaos*, 31(10):103102, 2021. (Cited on p. 87)

[218] W. Parker, R. Jenkins, C. Butler, and G. Abbott. Flash method of determining thermal diffusivity, heat capacity, and thermal conductivity. *Journal of Applied Physics*, 32(9):1679–1684, 1961. (Cited on p. 105)

[219] A. Paszke, S. Gross, F. Massa, A. Lerer, J. Bradbury, G. Chanan, T. Killeen, Z. Lin, N. Gimelshein, L. Antiga, A. Desmaison, A. Kopf, E. Yang, Z. DeVito, M. Raison, A. Tejani, S. Chilamkurthy, B. Steiner, L. Fang, J. Bai, and S. Chintala. PyTorch: An imperative style, high-performance deep learning library. In H. Wallach, H. Larochelle, A. Beygelzimer, F. d'Alché Buc, E. Fox, and R. Garnett, editors, *Advances in Neural Information Processing Systems 32*, pages 8024–8035. Curran Associates, Inc., 2019. (Cited on p. 96)

[220] D. Pedamonti. Comparison of non-linear activation functions for deep neural networks on MNIST classification task. *arXiv preprint arXiv:1804.02763*, 2018. (Cited on p. 103)

[221] M. Penwarden, S. Zhe, A. Narayan, and R. M. Kirby. A metalearning approach for physics-informed neural networks (PINNs): Application to parameterized PDEs. *Journal of Computational Physics*, 477:111912, 2023. (Cited on p. 107)

[222] C. A. Pickover. *The Pattern Book: Fractals, Art and Nature*. World Scientific, 1995. (Cited on p. 49)

[223] A. Pinkus. Approximation theory of the MLP model in neural networks. *Acta Numerica*, 8:143–195, 1999. (Cited on p. 91)

[224] A. Pinkus. Weierstrass and approximation theory. *Journal of Approximation Theory*, 107(1):1–66, 2000. (Cited on p. 47)

[225] H. Poincaré. *Les méthodes nouvelles de la mécanique céleste: Méthodes de MM. Newcomb, Glydén, Lindstedt et Bohlin. 1893*, volume 2. Gauthier-Villars it fils, 1893. (Cited on p. 42)

[226] M. Pollicott and M. Yuri. *Dynamical Systems and Ergodic Theory*, London Math. Soc. Stud. Texts 40. Cambridge University Press, 1998. (Cited on pp. 66, 82)

[227] S. Prajna. Barrier certificates for nonlinear model validation. *Automatica*, 42(1):117–126, 2006. (Cited on p. 87)

[228] C. Preston. What you need to know to knead. *Advances in Mathematics*, 78(2):192–252, 1989. (Cited on p. 139)

[229] J. L. Proctor, S. L. Brunton, and J. N. Kutz. Dynamic mode decomposition with control. *SIAM Journal on Applied Dynamical Systems*, 15(1):142–161, 2016. (Cited on p. 35)

[230] D. C. Psichogios and L. H. Ungar. A hybrid neural network-first principles approach to process modeling. *AIChE Journal*, 38(10):1499–1511, 1992. (Cited on p. 99)

[231] Q. Qu, J. Sun, and J. Wright. Finding a sparse vector in a subspace: Linear sparsity using alternating directions. In *Advances in Neural Information Processing Systems*, 27, 2014. (Cited on p. 60)

[232] J. Quandt. On the Hartman-Grobman theorem for maps. *Journal of Differential Equations*, 64(2):154–164, 1986. (Cited on p. 112)

[233] M. Raissi. Deep hidden physics models: Deep learning of nonlinear partial differential equations. *The Journal of Machine Learning Research*, 19(1):932–955, 2018. (Cited on p. 104)

[234] M. Raissi, P. Perdikaris, and G. E. Karniadakis. Physics-informed neural networks: A deep learning framework for solving forward and inverse problems involving nonlinear partial differential equations. *Journal of Computational Physics*, 378:686–707, 2019. (Cited on pp. 99, 104, 105)

[235] M. Raissi, A. Yazdani, and G. E. Karniadakis. Hidden fluid mechanics: Learning velocity and pressure fields from flow visualizations. *Science*, 367(6481):1026–1030, 2020. (Cited on p. 105)

[236] P. A. Reinbold, D. R. Gurevich, and R. O. Grigoriev. Using noisy or incomplete data to discover models of spatiotemporal dynamics. *Physical Review E*, 101(1):010203, 2020. (Cited on pp. 41, 61)

[237] P. A. Reinbold, L. M. Kageorge, M. F. Schatz, and R. O. Grigoriev. Robust learning from noisy, incomplete, high-dimensional experimental data via physically constrained symbolic regression. *Nature Communications*, 12(1):1–8, 2021. (Cited on pp. 41, 61)

[238] R. Rojas. The backpropagation algorithm. In *Neural Networks*, pages 149–182. Springer, 1996. (Cited on p. 95)

[239] F. J. Romeiras, C. Grebogi, E. Ott, and W. Dayawansa. Controlling chaotic dynamical systems. *Physica D: Nonlinear Phenomena*, 58(1-4):165–192, 1992. (Cited on p. 48)

[240] S. H. Rudy, S. L. Brunton, J. L. Proctor, and J. N. Kutz. Data-driven discovery of partial differential equations. *Science Advances*, 3(4):e1602614, 2017. (Cited on p. 62)

[241] L. Ruthotto and E. Haber. Deep neural networks motivated by partial differential equations. *Journal of Mathematical Imaging and Vision*, 62(3):352–364, 2020. (Cited on p. 107)

[242] F. Sakketou and N. Ampazis. On the invariance of the SELU activation function on algorithm and hyperparameter selection in neural network recommenders. In *IFIP International Conference on Artificial Intelligence Applications and Innovations*, pages 673–685. Springer, 2019. (Cited on p. 103)

[243] J. A. Sanders, F. Verhulst, and J. Murdock. *Averaging Methods in Nonlinear Dynamical Systems*, volume 59. Springer, 2007. (Cited on p. 54)

[244] H. Schaeffer and S. G. McCalla. Sparse model selection via integral terms. *Physical Review E*, 96(2):023302, 2017. (Cited on p. 41)

[245] C. Schlosser. Converging approximations of attractors via almost Lyapunov functions and semidefinite programming. *IEEE Control Systems Letters*, 6:2912–2917, 2022. (Cited on p. 87)

[246] C. Schlosser and M. Korda. Converging outer approximations to global attractors using semidefinite programming. *Automatica*, 134:109900, 2021. (Cited on p. 87)

[247] P. J. Schmid. Dynamic mode decomposition of numerical and experimental data. *Journal of Fluid Mechanics*, 656:5–28, 2010. (Cited on p. 11)

[248] P. J. Schmid. Dynamic mode decomposition and its variants. *Annual Review of Fluid Mechanics*, 54:225–254, 2022. (Cited on p. 11)

[249] O. T. Schmidt and T. Colonius. Guide to spectral proper orthogonal decomposition. *AIAA Journal*, 58(3):1023–1033, 2020. (Cited on p. 7)

[250] P. H. Schönemann. A generalized solution of the orthogonal procrustes problem. *Psychometrika*, 31(1):1–10, 1966. (Cited on p. 15)

[251] S. Schonsheck, J. Chen, and R. Lai. Chart auto-encoders for manifold structured data. *arXiv preprint arXiv:1912.10094*, 2019. (Cited on p. 139)

[252] S. C. Schonsheck, S. Mahan, T. Klock, A. Cloninger, and R. Lai. Semi-supervised manifold learning with complexity decoupled chart autoencoders. *arXiv preprint arXiv:2208.10570*, 2022. (Cited on p. 139)

[253] D. E. Shea, S. L. Brunton, and J. N. Kutz. SINDy-BVP: Sparse identification of nonlinear dynamics for boundary value problems. *Physical Review Research*, 3(2):023255, 2021. (Cited on p. 62)

[254] Y. Shin, J. Darbon, and G. E. Karniadakis. On the convergence of physics informed neural networks for linear second-order elliptic and parabolic type PDEs. *Communications in Computational Physics*, 28(5):2042–2074, 2020. (Cited on p. 107)

[255] A. Shoshitaishvili. Bifurcations of topological type at singular points of parametrized vector fields. *Functional Analysis and Its Applications*, 6(2):169–170, 1972. (Cited on p. 121)

[256] H. Simon. *Neural Networks: A Comprehensive Foundation*. Prentice Hall, 1999. (Cited on p. 91)

[257] E. D. Sontag. A Lyapunov-like characterization of asymptotic controllability. *SIAM Journal on Control and Optimization*, 21(3):462–471, 1983. (Cited on p. 78)

[258] E. D. Sontag. *Mathematical Control Theory: Deterministic Finite Dimensional Systems*, Texts in Appl. Math. 6. Springer Science & Business Media, 2013. (Cited on p. 78)

[259] J. Sprott. Simplest dissipative chaotic flow. *Physics Letters A*, 228(4-5):271–274, 1997. (Cited on p. 52)

[260] B. Stevens and T. Colonius. FiniteNet: A fully convolutional LSTM network architecture for time-dependent partial differential equations. *arXiv preprint arXiv:2002.03014*, 2020. (Cited on p. 107)

[261] W. A. Strauss. *Partial Differential Equations: An Introduction*. John Wiley, 2007. (Cited on p. 5)

[262] S. Streif, D. Henrion, and R. Findeisen. Probabilistic and set-based model invalidation and estimation using LMIs. *IFAC Proceedings Volumes*, 19(3):4110–4115, 2014. (Cited on p. 87)

[263] S. Streif, P. Rumschinski, D. Henrion, and R. Findeisen. Estimation of consistent parameter sets for continuous-time nonlinear systems using occupation measures and LMI relaxations. In *Proceedings of the 52nd IEEE Conference on Decision & Control*, pages 6379–6384, 2013. (Cited on p. 87)

[264] S. H. Strogatz. *Nonlinear Dynamics and Chaos: With Applications to Physics, Biology, Chemistry, and Engineering*. CRC Press, 2018. (Cited on pp. 3, 44)

[265] W. Su, M. Bogdan, and E. Candes. False discoveries occur early on the lasso path. *The Annals of Statistics*, 45(5):2133–2150, 2017. (Cited on p. 38)

[266] Y. Sun, L. Zhang, and H. Schaeffer. NeuPDE: Neural network based ordinary and partial differential equations for modeling time-dependent data. In *Mathematical and Scientific Machine Learning*, pages 352–372. PMLR, 2020. (Cited on p. 107)

[267] N. E. Taghvayi and B. Aazhang. Application of embedded dynamic mode decomposition on epileptic data for seizure prediction. In *2019 53rd Asilomar Conference on Signals, Systems, and Computers*, pages 1705–1708. IEEE, 2019. (Cited on p. 21)

[268] N. Takeishi, Y. Kawahara, and T. Yairi. Learning Koopman invariant subspaces for dynamic mode decomposition. In *Advances in Neural Information Processing Systems*, 30, 2017. (Cited on p. 21)

[269] F. Takens. Detecting strange attractors in turbulence. In *Dynamical Systems and Turbulence, Warwick 1980*, pages 366–381. Springer, 1981. (Cited on p. 21)

[270] W. Tan, U. Topcu, P. Seiler, G. Balas, and A. Packard. Simulation-aided reachability and local gain analysis for nonlinear dynamical systems. In *2008 47th IEEE Conference on Decision and Control*, pages 4097–4102. IEEE, 2008. (Cited on p. 78)

[271] M. Tao, H. Owhadi, and J. E. Marsden. Nonintrusive and structure preserving multiscale integration of stiff ODEs, SDEs, and Hamiltonian systems with hidden slow dynamics via flow averaging. *Multiscale Modeling & Simulation*, 8(4):1269–1324, 2010. (Cited on p. 57)

[272] R. Tibshirani. Regression shrinkage and selection via the lasso. *Journal of the Royal Statistical Society: Series B (Methodological)*, 58(1):267–288, 1996. (Cited on p. 38)

[273] I. Tobasco, D. Goluskin, and C. R. Doering. Optimal bounds and extremal trajectories for time averages in nonlinear dynamical systems. *Physics Letters A*, 382(6):382–386, 2018. (Cited on pp. 66, 67)

[274] A. Tsantekidis, N. Passalis, A. Tefas, J. Kanniainen, M. Gabbouj, and A. Iosifidis. Forecasting stock prices from the limit order book using convolutional neural networks. In *2017 IEEE 19th Conference on Business Informatics (CBI)*, volume 1, pages 7–12. IEEE, 2017. (Cited on p. 107)

[275] W. Tucker. The Lorenz attractor exists. *Comptes Rendus de l'Académie des Sciences-Series I-Mathematics*, 328(12):1197–1202, 1999. (Cited on p. 5)

[276] S. M. Ulam. *A Collection of Mathematical Problems*. Interscience Tracts in Pure and Applied Mathematics 8. Interscience Publishers, 1960. (Cited on p. 87)

[277] G. Valmorbida, M. Ahmadi, and A. Papachristodoulou. Stability analysis for a class of partial differential equations via semidefinite programming. *IEEE Transactions on Automatic Control*, 61(6):1649–1654, 2015. (Cited on p. 87)

[278] M. V. Valueva, N. Nagornov, P. A. Lyakhov, G. V. Valuev, and N. I. Chervyakov. Application of the residue number system to reduce hardware costs of the convolutional neural network implementation. *Mathematics and Computers in Simulation*, 177:232–243, 2020. (Cited on p. 107)

[279] A. Van den Oord, S. Dieleman, and B. Schrauwen. Deep content-based music recommendation. In *Advances in Neural Information Processing Systems*, 26, 2013. (Cited on p. 107)

[280] B. Van der Pol. On "relaxation-oscillations." *The London, Edinburgh, and Dublin Philosophical Magazine and Journal of Science*, 2(11):978–992, 1926. (Cited on p. 22)

[281] E. Vasconcelos Filho and P. L. dos Santos. A dynamic mode decomposition approach with Hankel blocks to forecast multi-channel temporal series. *IEEE Control Systems Letters*, 3(3):739–744, 2019. (Cited on p. 21)

[282] W. Walter. *Ordinary Differential Equations*, Grad. Texts in Math. 182. Springer Science & Business Media, 2013. (Cited on p. 2)

[283] P. Walters. *An Introduction to Ergodic Theory*, Grad. Texts in Math. 79. Springer Science & Business Media, 2000. (Cited on p. 82)

[284] M. Wanner and I. Mezić. Robust approximation of the stochastic Koopman operator. *SIAM Journal on Applied Dynamical Systems*, 21(3):1930–1951, 2022. (Cited on p. 74)

[285] M. O. Williams, I. G. Kevrekidis, and C. W. Rowley. A data-driven approximation of the Koopman operator: Extending dynamic mode decomposition. *Journal of Nonlinear Science*, 25(6):1307–1346, 2015. (Cited on pp. 26, 28)

[286] M. O. Williams, C. W. Rowley, and I. G. Kevrekidis. A kernel-based method for data-driven Koopman spectral analysis. *Journal of Computational Dynamics*, 2(2):247–265, 2015. (Cited on pp. 29, 31)

[287] A. Wolf, J. B. Swift, H. L. Swinney, and J. A. Vastano. Determining Lyapunov exponents from a time series. *Physica D: Nonlinear Phenomena*, 16(3):285–317, 1985. (Cited on p. 136)

[288] H. Wolkowicz, R. Saigal, and L. Vandenberghe. *Handbook of Semidefinite Programming: Theory, Algorithms, and Applications*, Internat. Ser. Oper. Res. Manag. Sci. 27. Springer Science & Business Media, 2012. (Cited on p. 147)

[289] C. Wunsch. The long-period tides. *Reviews of Geophysics*, 5(4):447–475, 1967. (Cited on p. 16)

[290] C. Wunsch, D. B. Haidvogel, M. Iskandarani, and R. Hughes. Dynamics of the long-period tides. *Progress in Oceanography*, 40(1-4):81–108, 1997. (Cited on p. 16)

[291] H. Xu and D. Zhang. Robust discovery of partial differential equations in complex situations. *Physical Review Research*, 3(3):033270, 2021. (Cited on p. 106)

[292] G. Yalnız, B. Hof, and N. B. Budanur. Coarse graining the state space of a turbulent flow using periodic orbits. *Physical Review Letters*, 126(24):244502, 2021. (Cited on p. 138)

[293] D. Yang, Z. Ma, and A. Buja. A sparse singular value decomposition method for high-dimensional data. *Journal of Computational and Graphical Statistics*, 23(4):923–942, 2014. (Cited on p. 60)

[294] W. Zha, W. Zhang, D. Li, Y. Xing, L. He, and J. Tan. Convolution-based model-solving method for three-dimensional, unsteady, partial differential equations. *Neural Computation*, 34(2):518–540, 2022. (Cited on p. 107)

[295] L. Zhang and H. Schaeffer. On the convergence of the SINDy algorithm. *Multiscale Modeling & Simulation*, 17(3):948–972, 2019. (Cited on p. 39)

[296] Q. Zhang, Y. Chen, and Z. Yang. Data-driven solutions and discoveries in mechanics using physics informed neural network. *Preprints*, 2020060258, 2020. (Cited on p. 105)

[297] Y. Zheng, G. Fantuzzi, and A. Papachristodoulou. Sparse sum-of-squares (SOS) optimization: A bridge between DSOS/SDSOS and SOS optimization for sparse polynomials. In *2019 American Control Conference (ACC)*, pages 5513–5518. IEEE, 2019. (Cited on p. 88)

[298] Y. Zhu, N. Zabaras, P.-S. Koutsourelakis, and P. Perdikaris. Physics-constrained deep learning for high-dimensional surrogate modeling and uncertainty quantification without labeled data. *Journal of Computational Physics*, 394:56–81, 2019. (Cited on p. 107)

Index